870403

GEOGRAPHIC ESSENTIALS
B.C. edition

GEOGRAPHIC ESSENTIALS

Map Skills Using the Canadian Oxford School Atlas 5th Edition
B.C.

Walter G. Kemball

Toronto
Oxford University Press

© Oxford University Press (Canada) 1986

All rights reserved — no part of this book may be reproduced in any form without the written permission of the publisher. Photocopying or reproducing mechanically in any other way parts of this book without permission in writing from the publisher is an infringement of the copyright law.

Illustrations by Fred Huffman

ISBN 0 19 540559-5

Printed in Canada by
Hemlock Printers Ltd.
Burnaby, B.C.

2 3 4 5 6 7 8 9   92 91 90 89 88 87

CONTENTS

PART 1: Basic Skills
MAPS: AN INTRODUCTION/1

MAP BASICS/4
- Title/4
- Legend/4
- Scale/6

OUTLINE MAPS/8
- Shapes/8

SYMBOLS/12

GRIDS/15
- Plotting Position/18
- Plotting a Point/20

PART 2: Map and Atlas Skills
DISTRIBUTION/22

COLOUR ON ATLAS MAPS/24
- Relief Shading/24
- Other Uses of Colour/26

SYMBOLS ON ATLAS MAPS/27

DIRECTION/30

LATITUDE AND LONGITUDE/32
- Parallels of Latitude/33
- Meridians of Longitude/39
- The Earth Grid/42

GAZETTEER/46

TIME ZONES/48

SCALE/55
- Map Scale/56
- Scales of Atlas Maps/60

TYPES OF MAPS USED IN THE CANADIAN OXFORD SCHOOL ATLAS, 5TH EDITION/66
- Place Name or Political Maps/66
- Thematic Maps/68
- Physical Maps/69
- Regional Topographic Maps/70

MAP PROJECTIONS/72
  Map Projections Used in the Canadian Oxford School Atlas, 5th Edition/75
    Conical Orthomorphic/76
    Cylindrical/77
    Zenithal Equidistant/80
    Zenithal Equal Area/81
GREAT CIRCLE ROUTES/82
COMPASS BEARINGS/84
  Compass Bearings on a Mercator Map/85

PART 3: Applied Skills

PHOTOGRAPHS/87
  Patterns on Photographs/87
  Types of Photographs/87
    Air Photographs/89
      Oblique Air Photographs/89
      Vertical Air Photographs/92
SIZE AND LOCATION/95
  Canada's Location/95
  Exact Position/96
  Size/97
  Boundaries/97
PROFILES AND CROSS SECTIONS/99
  Profiles/99
  Cross Sections/100
PHYSICAL FEATURES/101
  Landforms/101
    Ancient Shields/101
    Sedimentary Rocks/102
    Uplifted Remains of Ancient Mountain Systems/103
    Younger Fold Mountains/104
    Recent Deposits/105
    River Systems/108
    Patterns made by Rivers/108
    Drainage Basins/109
    Spillways/110
CLIMATE/113
  Temperature Maps/113
  High and Low Pressure Systems/116
  Ocean Currents/117
    Climate Controls that Affect Temperature/118
  Mean Annual Precipitation Maps/123

    Climate Controls that Affect Precipitation/124
  Climate Graphs/128
    Line Graphs/129
    Bar Graphs/132
    Interpretation of Climate Graphs/136
    Climate Controls for Edmonton/138
    Climate Regions/140

NATURAL VEGETATION/141

AGRICULTURE/146
  Factors that Affect Farming/146
  Wheat Farming/149
  Rice Farming/150

POPULATION/152
  Population Density/155

PART 4: Historical Applications

HUMAN SETTLEMENT PATTERNS/159
  Reasons for the Location of Settlements/162
  Settlement Growth/163

TRAVEL/170

GLOSSARY/177

INDEX/182

ACKNOWLEDGEMENTS/187

PART 1:
Basic Skills

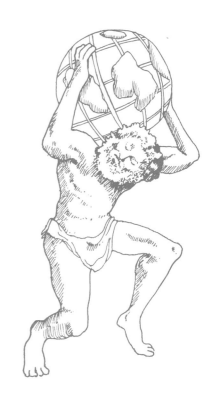

MAPS: AN INTRODUCTION

Every day we hear on radio and television about events taking place in other parts of Canada and the world. We read in newspapers and books about exotic foreign places. Each year more people travel and they want to learn more about where they are going, and how to get there. Atlases bring this information to them.

An atlas is a collection of maps bound together into a book. In Greek mythology, Atlas was punished for taking part in a rebellion by being made to hold up the sky with his head and hands. Eventually people thought that Atlas was supporting the world, and his picture was placed on maps. The first person to use the word "atlas" to mean a group of maps was the Flemish geographer Mercator, in a book published in 1595.

The first atlases contained maps that pictured the world as it was then known.

Ptolemy, a Greek who lived in the second century, made some of the most famous and accurate maps. This is a map of the world as it was known in 150 A.D., showing Europe, Asia, and parts of Africa.

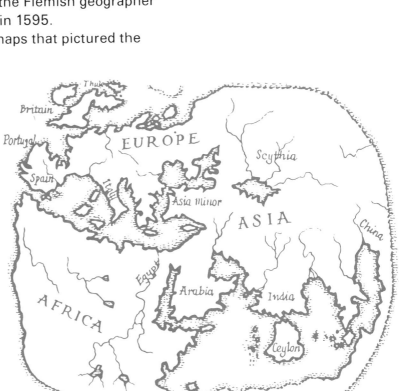

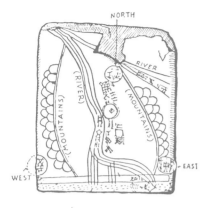

The world's oldest map. A clay tablet dated about 2500 B.C. showing part of Mesopotamia.

Champlain's map of New France, 1632.

General atlases still give a picture of the whole world, but they use a great variety of maps to do so. Some atlas maps show the whole world; others show only part of the world. These are the kinds of maps used by most people for general information. Some maps show one topic in detail, for example, the map showing the population of Canada on page 10 of the *Canadian Oxford School Atlas* (5th edition). Such a map is called a thematic map since it shows only one theme or topic.

If we tried to put all the information we have on one map, the map would be too crowded and we would not be able to read it. That is why atlases have a variety of maps. Some atlases also include other resources in order to provide information that cannot be shown on maps. The *Canadian Oxford School Atlas* includes statistics, graphs, charts, and tables. In other atlases diagrams, pictures, and photographs are used to provide more detail.

Some atlases show only parts of the earth: the oceans *(Atlas of the Oceans)*, some provinces *(Atlas of the Prairie Provinces)*. Many specialized atlases deal with themes such as the heavens *(Atlas of the Universe)*, or economics *(Oxford Regional Economic Atlas of the United States and Canada)*.

MAP BASICS

ACTIVITY 1

In your notebook draw a chart similar to the one below.
   Examine the maps on pp. 20-21; 76; 136-137; and 144 of the *Canadian Oxford School Atlas* (5th edition). List the items that all these maps have in common.

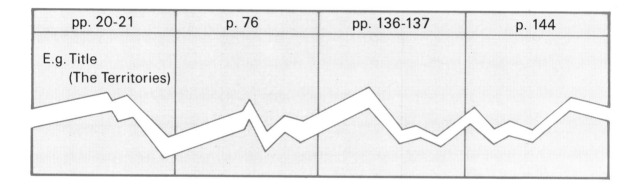

pp. 20-21	p. 76	pp. 136-137	p. 144
E.g. Title (The Territories)			

The items you have listed are known as map basics.

TITLE

Many people use a road map when they go for a long drive or on a touring holiday. The first basic you would look for on a road map is the title. This shows whether the map is of the area in which you will be driving. The title usually gives the name of one province (British Columbia) or a group of provinces (Atlantic Provinces).

LEGEND

Secondly, you would examine the legend of the map to see what kind of information the map provides.

ACTIVITY 2

Look at the road map legend to answer the following questions.

1. Which one of the major sections would be of most help in planning your route?
2. What information in the legend would help you decide what you wanted to see on your drive?
3. What service is provided for tourists who want more information about the areas in which they are driving?
4. If you were on a camping trip, what information would be particularly important?
5. What would be of particular importance if you were camping and had an allergy to insect bites?
6. You do not want to use busy highways but would prefer to drive on country roads. What section of the legend would be of help in planning your route?

SCALE

The third map basic is scale. One advantage of a road map is that it gives three systems of finding distances. The first is to add the kilometre distances between red bullets: this will give you the total road distance.

Another system is to use the map scale to measure the distances between places. How the map scale works will be explained later in Part 2.

The third system of finding distances is to use the kilometric distances chart, which gives you the distance between major centres. For example, to find the

Distance

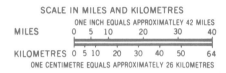

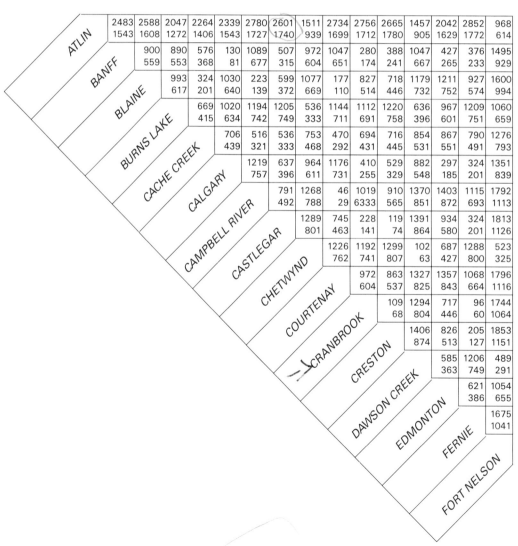

distance between Banff and Creston you read across from Banff and up from Creston. The number 388, where the columns meet, is the distance in kilometres.

ACTIVITY 3
1. How would you use the map scale to decide how long your drive will take?
2. Which of the three methods of showing distances would you use to find:
    (a) the shortest distance between two places?
    (b) the road distance between small communities?
    (c) the road distance between large communities?
3. Use the chart to find the distance between:
    (a) Atlin and Castlegar
    (b) Burns Lake and Fernie
    (c) Cache Creek and Calgary

OUTLINE MAPS

In science, you use a notebook and diagrams to record information about experiments. In geography, outline maps are often used to record information about areas you are studying.

A peninsula is a piece of land jutting out into the water. Which is the peninsula on this map?

When you use an outline map it is important that you recognize the difference between land and water areas.

If you had drawn this map, how could you tell someone which is land and which is water?

Whenever you work with an outline map, make sure you know which areas are land and which are water.

BLACK PENINSULA

ACTIVITY 4
1. If you were to colour an outline map of Canada to separate water and land areas, which colours might you use? Why is it easier to work with a coloured map than a black and white outline map?
2. Use your atlas to help place the following basics on an outline map of Canada.
   (a) title and subtitle
   (b) names of provinces and territories
   (c) provincial capitals (use the same symbol for each capital) and the federal capital (use a different symbol)
   (d) legend (show each symbol and what it means, and boundaries)

Shapes

When you examine a map, the first thing that may catch your eye is the distinctive shape of such features as continents, islands, or bodies of water. After you have used many maps you will immediately recognize some of these features. Their shapes will be the same on maps at different scales. Some maps are not oriented to the north, but you may still recognize the shape of a feature even when it is "upside-down."

To remember specific shapes, it is often helpful to compare them with geometric figures — Antarctica is like a circle, North America is like a triangle; or with objects — Italy is shaped like a boot, New Guinea looks like an animal.

ACTIVITY 5
1. Refer to the map on pp. 2-3 of the atlas to name these outlines of islands and provinces (next page) of Canada.

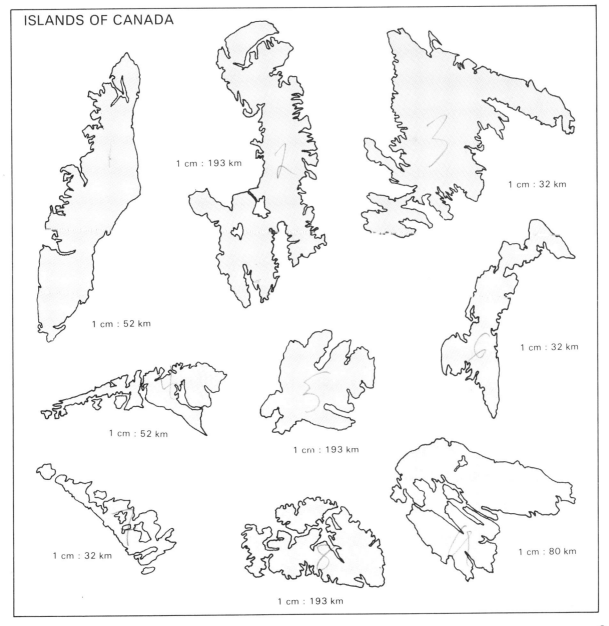

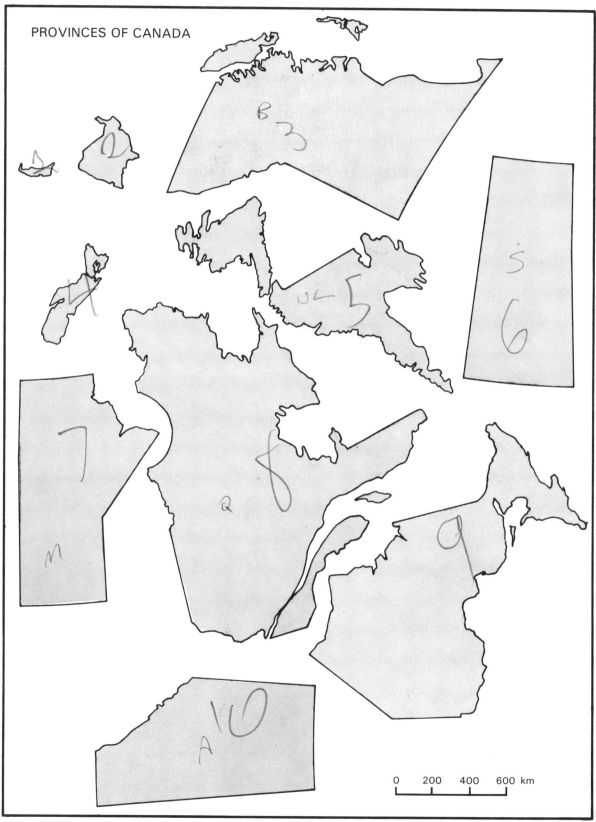

ACTIVITY 6
1. Use the outlines below to name each continent and describe its shape.
2. Use copies of these shapes to create your own world map by placing them in the correct relative positions.

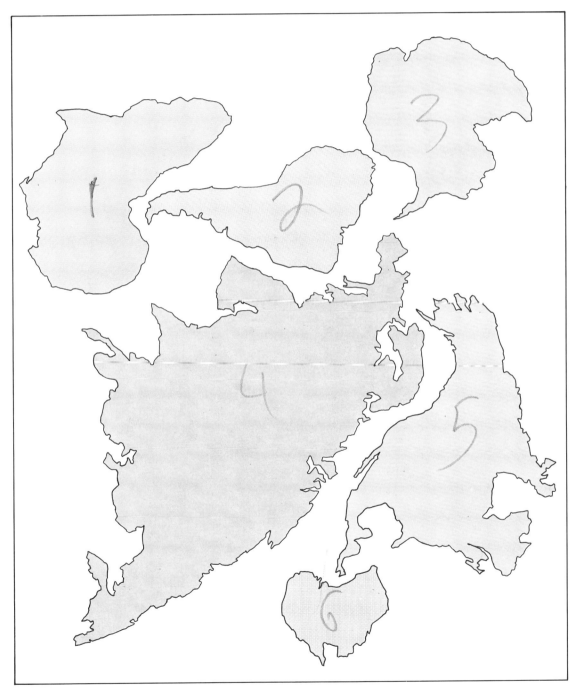

SYMBOLS

We live in a world of symbols. Symbols are a kind of shorthand. They are useful because people can recognize them easily — even people who speak different languages. What do these symbols mean?

Keep a class list of symbols in everyday use. Add to it as you find new ones.

Map symbols are also a type of shorthand. Map symbols are marks, lines, and colours on a map that stand for real things on the surface of the earth. Sometimes the marks are pictures that suggest the things they represent.

These are called pictorial symbols. A map that has pictorial symbols is called a pictorial map.

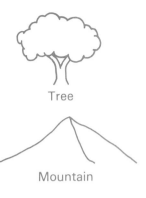
Tree

Mountain

Hill

ACTIVITY 7
1. What does each symbol on the pictorial map of Australia represent?
2. What general information about Australia do you get from looking at this map?
3. Draw a pictorial map of your school area. Make up a symbol for each object you show on the map. Make sure your completed map contains all the important map basics.

If you had tried to show every house on the pictorial map of your school area, you would not have had enough room. Rather than drawing a picture of a house or a school, it would be simpler and would save space if you used a symbol.

ACTIVITY 8
Refer to the map legend on page 5 of this book.
1. Which symbol is used to show:
    (a) Yellowhead Highway;
    (b) railway?
2. Which symbol is used to show:
    (a) fish hatchery;
    (b) information centre?
3. What things are shown by line symbols?
4. Draw the symbol for each of the following:
    (a) Trans-Canada Highway;
    (b) Interstate Highway;
    (c) airport;
    (d) port of entry;
    (e) ferry route.

GRIDS

How would you give the location of the circle in this area?

You might say that it is in the lower left corner, but this is not very accurate. One way to overcome this problem is to use a grid.

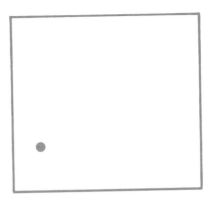

Draw a grid similar to the one on the right in your notebook.

Start at the lower left hand corner, and number along the bottom line from left to right. Then put letters along the side from bottom to top. The location of a point is given with the number first, followed by the letter.

Now, if you were asked to give the location of the circle, what grid reference would you give?

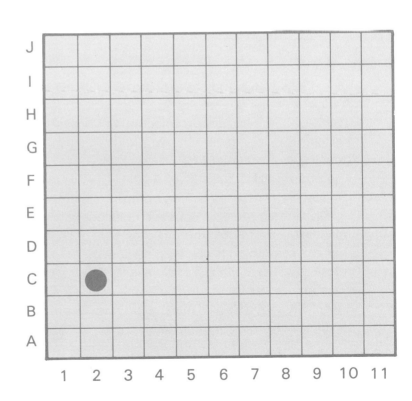

ACTIVITY 9
1. Give the grid location of each triangle on the diagram at the right.
2. Put an X on each of the following in the grid you have drawn:
   (a) 2 I
   (b) 6 F
   (c) 8 C
   (d) 1 J

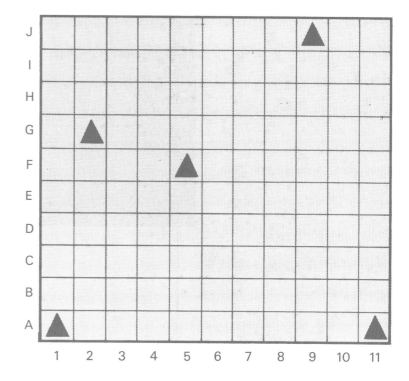

3. Play a game of battleship.
   (a) Each player has a grid sheet (16 x 16 squares) and draws a number of ships on it.
   1 battleship   3 squares (horizontal or vertical)
   1 cruiser   2 squares (horizontal or vertical)
   1 destroyer   1 square
   1 submarine   1 square
   (b) The players take turns calling out grid references, e.g. 4 G. When the reference includes a whole ship, that ship is sunk. If it includes only part of a ship, the remaining parts must be called out before the ship is sunk. Each player tells whether it is a hit or miss when grid references are given.
   (c) Both players keep track of their calls (O) and the other player's calls (X) on their cards.
   (d) The first player to sink the other's ships is the winner.
   (e) The game can be made more difficult by increasing the number of grid spaces and by adding more ships.

4. Use the map below to give the grid references for the following:
   (a) High Bluff
   (b) Bear Hollow
   (c) Dune View
   (d) the airport
   (e) the yacht club
   (f) the golf club

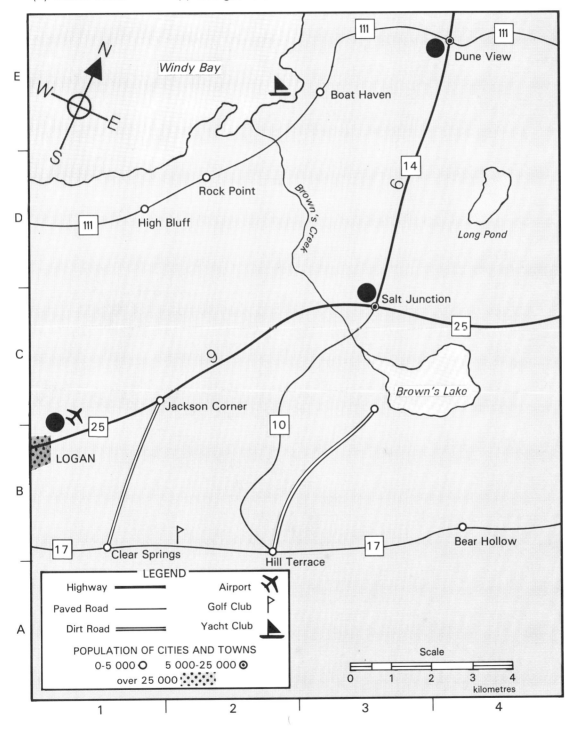

Plotting Position

Another way to plot position is to use two numbers instead of a number/letter combination.

The following diagram shows the parking spots in a company parking lot. The occupied spots have the names of the people using them. Vacant spots are marked with a "V."

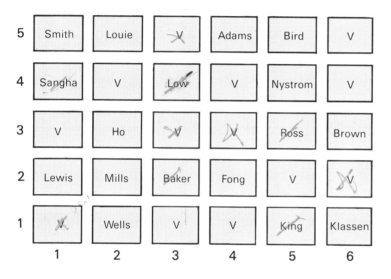

The location of a parking spot is given by the position across (bottom number), then the position up (side number). Put a comma between the numbers. For example, the Ross parking spot is at 5,3.

ACTIVITY 10
1. Who is parked in the following positions?
   (a) 2,2 (e) 1,5
   (b) 4,5 (f) 6,3
   (c) 6,1 (g) 4,2
   (d) 2,3 (h) 5,5
2. Give the position of the spot occupied by these people.
   (a) Louie   (d) Nystrom
   (b) King   (e) Wells
   (c) Sangha (f) Low
3. Some cars were taken out as salespeople went out on calls. Cover the name on the spot they had with a little piece of paper.
   (a) 3,4 (d) 5,1
   (b) 5,3 (e) 3,2
   (c) 1,4
4. During the day, other cars arrived and were given spots as follows.
   (a) Anderson 1,1 (d) Goldstein 6,2
   (b) Duff 3,3   (e) Fulton 4,3
   (c) Chou 3,5
   Print the names on small pieces of paper and place them on the right spots.
5. How many cars were in the lot after the new arrivals had parked?
6. One row across now has only one empty spot. Give the position of this empty spot.
7. Give the position of the following empty spots.
   (a) between Mills and Fong
   (b) between Ho and Louie
   (c) between Fulton and Adams

Plotting a Point

In the diagram below, the complete set of small squares is called a grid.

The bottom numbered line (horizontal axis) is called the "eastings." The side numbered line (vertical axis) is called the "northings."

To plot the position of a point, give the eastings (number across) first, and the northings (number up) second. A plotted point is at an "intersection" where the eastings line from the bottom crosses the northings line from the side. For example, the upper end of the topmost line in the diagram is at position 4,13.

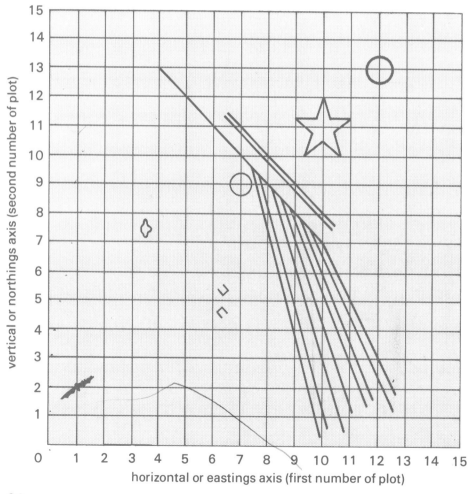

ACTIVITY 11
1. Draw a grid like the one on p. 20 (or use grid paper). Plot these points on the grid and letter each one.

A — 7,0	K — 4,5
B — 8,2	L — 3,6
C — 9,4	M — 4,8
D — 8,3	N — 6,9
E — 7,3	O — 5,10
F — 6,4	P — 6,11
G — 7,5	Q — 14,15
H — 7,6	R — 10,7
I — 6,5	S — 12,5
J — 5,6	

2. Join the points with straight lines using a ruler and pencil. What is the drawing?
3. Give the grid positions for the following points.
   (a) centre of the eye
   (b) body of the bat
   (c) centre of the star
   (d) centre of the circle
   (e) top point of the star

PART 2:
Map and Atlas Skills

DISTRIBUTION

Distribution refers to the way things are spread over the land surface. The way things are positioned may be either:
- (a) thought about, planned and deliberately done;
  OR
- (b) simple chance, not planned, or owing to an act of nature

A. This is a planned distribution.

B. This is an act of nature. The cows may go anywhere in the field.

C. This distribution is not planned.

ACTIVITY 12
Note whether the following distributions are:
    P — planned;
    NP — not planned.

1. snowfall
2. houses in a town
3. ash from a volcano
4. floats in a parade
5. fruit trees in an orchard
6. a swarm of bees
7. people sunning on a beach
8. milk spilled on the floor
9. debris from a hurricane

All atlas maps show distributions through the use of colours or symbols.

ACTIVITY 13
Refer to the following maps in the atlas and indicate if the distributions shown are planned or not planned.
1. Physiographic Regions, p. 8
2. Population, p. 10
3. Transportation, p. 11
4. Manufacturing, p. 18
5. Chicago, p. 73
6. Vegetation, p. 145

COLOUR ON ATLAS MAPS

RELIEF SHADING

Atlas maps use colour to present information about distributions.

ACTIVITY 14
Examine the colour legend on p. 133 of the atlas. This method of colouring a map is known as "layer tinting" or "altitude tinting."

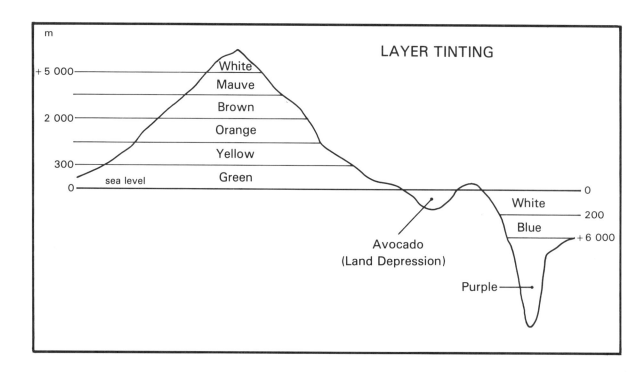

1. What do the colours represent?

2. Copy and complete the following chart.

DESCRIPTION	ELEVATIONS (m) (altitude above sea level*)	COLOUR
Highlands		brown, mauve and white
Middle lands	300 — 2 000	
Low lands		green
Land depression	sea level — variable	
Shallow depths	sea level — 200	
Middle depths	200 — 6 000	
Ocean deeps		purple

*sea level   the average level of the sea between high and low tides

3. How does the layer-tint method of colouring a map help you to visualize the land?

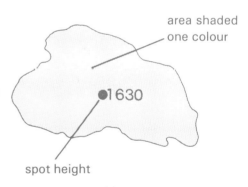

Numbers such as the one showing the depth of a depression or the height of a mountain are called spot heights. They show the exact height of the land within the colour-shaded area.

4. (a) Find a land depression area in Egypt on the atlas map on p. 129. What number beside this area shows the depth below sea level of this depression?
   (b) Keep a class list of land depression areas found anywhere in the world.

5. On the map on p. 20 of the atlas find and list the spot heights for Keele Pk., Mt. Campbell, Mt. Michelson and Mt. Sir James McBrien.

Metal markers such as this are set in the ground as base points for surveying the land. Sometimes these markers show the exact altitude of the ground as well.

OTHER USES OF COLOUR

You will notice as you use the atlas that most maps are layer tinted to show land altitude and water depth. However, colours are used on some maps to show other kinds of information.

ACTIVITY 15
Find the various uses for the colour blue on the following: atlas pp. 4 inset, 18, 100, 128-129 and most other maps.

   Subdivisions within a major region will usually be shown by various shades of the same colour. A good example of this is found on atlas p. 8 in the physiographic regions map. The interior plains are shown by two shades of mustard: a darker shade which represents the hills and plateaus, and a lighter shade which represents the lowlands.

SYMBOLS ON ATLAS MAPS

Symbols on atlas maps are generally of two main types:
   (a) line symbols such as those used for railways, roads and boundaries, and
   (b) mark symbols such as those used for airports, cities, and salt pans.

   Both types of symbols can represent natural features or features made by people.

ACTIVITY 16
1. Draw a chart like the one below in your notebook. Use the legend on p. 26 of the atlas to complete the chart.

OBJECT	SYMBOL	LINE OR MARK (check one)	MADE BY PEOPLE	NATURAL FEATURES

2. (a) What kinds of objects are represented by line symbols?
   (b) How can a line symbol be changed to show different information?
3. What kinds of objects are represented by mark symbols?
4. Suggest reasons why symbols are coloured differently.
5. What colours are used for symbols relating to water; air and road transportation; railways?
6. Sometimes a particular symbol is used so often to represent a feature that it is not even explained on map legends. Draw and give the name for one blue-line and one blue-mark symbol which are not shown in the legend.

When you study a number of legends (atlas pp. 21, 23, 130) you will notice that the same symbols are always used to show the same object. These are known as "conventional symbols" because they are generally accepted as the standard symbol for a particular object. However, some maps may have other symbols to show the same object. Since most maps have a legend, it is not necessary to memorize the symbols. What is necessary is to study the legend before trying to read a map.

Another method of presenting information through symbols is to vary the size of the symbol.

ACTIVITY 17
1. Examine the legend for the manufacturing map on atlas p. 19.
    (a) What do the various sizes of circles show?
    (b) What advantages can you suggest for this method of conveying information?
    (c) What has been done to the symbols to show different quantities of production?
    (d) If only the 3 largest circles had been used to show manufacturing, how many circles would there be on each map? How does this method of providing information affect the accuracy of what you see?

Line symbols may also be shown in different ways to convey information about quantity.

ACTIVITY 18
1. What is shown by varying the width of the shipping line symbol on p. 35 of the atlas, "Canadian Wheat Exports"?
2. On the Pacific Ocean map on atlas pp. 132-133:
    (a) How is the current symbol changed to show the difference between warm and cold currents?
    (b) Explain how the size of the current is shown by these symbols.
3. Explain how the regularity of the winds is shown on the maps on atlas pp. 142-143.

Population is often shown on maps. You could use a pictorial symbol:

Another way is to use a symbol to show each person, like this:

Still another way to represent them would be like this:

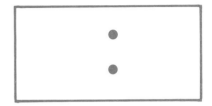

1 dot represents 6 people

ACTIVITY 19
1. How many people are represented here?

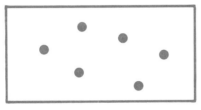

1 dot represents 10 people

2. (a) In which part of this area do most people live?
   (b) How many people live in this area?

1 dot represents 100 people

3. Draw a diagram to show the following:
   (a) total population 150 000
   (b) the population is split into three main groups:
       — 90 000 live in the north
       — 40 000 in the south
       — 20 000 in the west
4. Draw a map using symbols and colour to show the numbers and distribution of boys and girls in the classroom.
5. The population map of Canada on atlas p. 10 uses black symbols to represent numbers of people. How many people are represented by each of the three black symbols on this map?

DIRECTION

Many maps are drawn with north at the top. An arrow indicates direction.

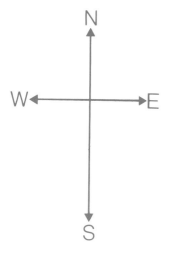

If north is not at the top, the arrow may show this:

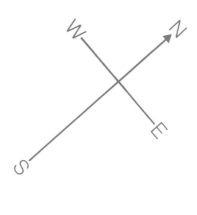

or this:

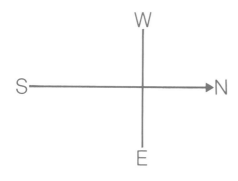

Sometimes when you travel from one place to another you will not travel directly north or south, east or west. The diagram shows these in-between directions.

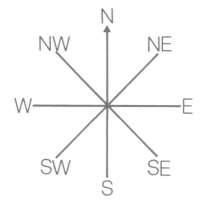

ACTIVITY 20

1. Copy the diagram and place an E at the end of the arrow that points east.

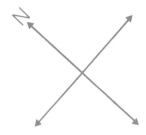

2. Copy the diagram and place a W at the end of the arrow that points west.

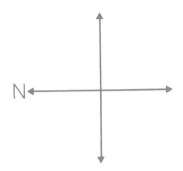

3. In this diagram, A is what direction from B? C is what direction from D?

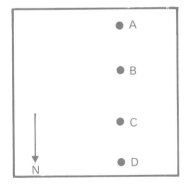

4. In this diagram what is the direction of B from A?

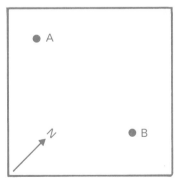

LATITUDE AND LONGITUDE

When you want to travel from one place to another on land, you can use road maps, landmarks, and signs to help find the way. A ship's captain in the middle of the ocean also has to find his way from one port to another. He has none of these signs to help him. Instead, he uses a grid made up of two sets of imaginary lines on the globe.

The horizontal lines of the grid are called parallels of latitude, and the vertical lines are known as meridians of longitude. Together they form the earth grid.

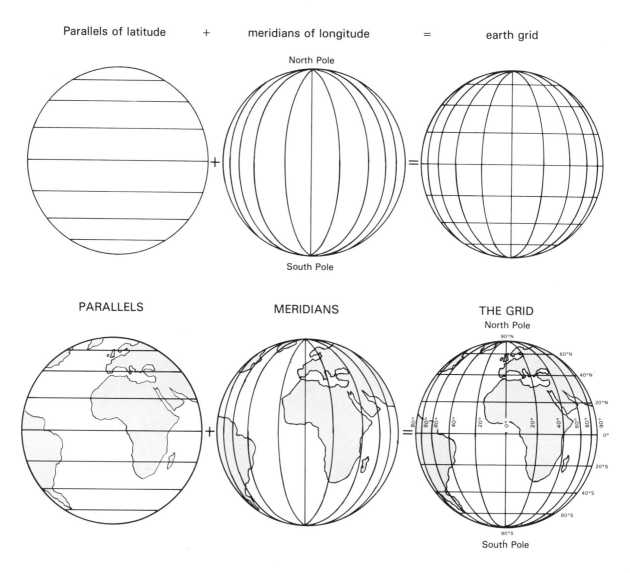

Parallels of Latitude

The line of latitude that circles the earth from west to east and divides it into two halves is the Equator (from a Latin word meaning "to make equal"). The Equator

divides the earth into two hemispheres (half spheres) known as the Northern Hemisphere and the Southern Hemisphere. Other imaginary lines running in an east-west direction are known as parallels because they are parallel to the Equator and to each other.

Parallels of latitude are numbered in degrees. A degree is a unit used to measure an angle. Parallels of latitude are measured by the angle they make with the equator at the centre of the earth (see diagram). Since the equator does not make an angle with itself, it is numbered 0°. The Equator is the starting line for numbering the parallels. There are 90 parallels between the Equator and the North Pole at 90°N, and 90 parallels

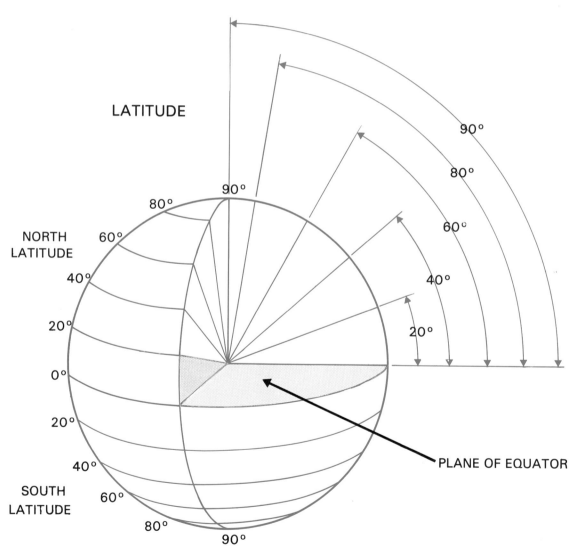

between the Equator and the South Pole at 90°S. The parallels are 1 degree apart (111 km).

A map will not show all the parallels because it would be too crowded. How many degrees apart are the parallels on the maps on atlas pp. 4-5, 40-41, and 140-141?

ACTIVITY 21

1. The Northern Hemisphere is located between the Equator and what point?
2. To what place would you travel to reach the most southerly point on earth?
3. What happens to the length of parallels of latitude as you move south from the North Pole to the Equator?
4. Since 1° of latitude is equal to approximately 111 km, how many kilometres is it from (a) the Equator to the North Pole, (b) the Equator to your home, and (c) your home to the North Pole?

Examine the map on atlas pp. 140-141 to answer the following.

5. In which hemisphere is more than half of the earth's land located?
6. Which two continents are located completely within the Southern Hemisphere?
7. Which two continents are split between the Southern and Northern Hemispheres?
8. Copy the diagram in your notebook and complete the numbering of the parallels of latitude.
9. State the latitude of A, B, and C.
10. Place a small circle at the following: 15°N, 5°S, 18°S
11. How many kilometres are there between A and C?

12. Refer to the map on pp. 22-23 of the atlas and give the latitude of the following:
    (a) Winnipeg, Manitoba
    (b) Edmonton, Alberta
    (c) Vancouver, British Columbia
13. Refer to the map on atlas pp. 110-111 to give the latitude of the following:
    (a) Tokyo, Japan
    (b) Calcutta, India
    (c) Kabul, Afghanistan
    (d) Ulan Bator, Mongolia
14. Refer to the map on atlas pp. 106-107 to give the latitude of the following:
    (a) Cairo, Egypt
    (b) Tehran, Iran
    (c) Ankara, Turkey
    (d) Karachi, Pakistan

Special Parallels of Latitude

On world maps four special parallels of latitude are usually shown by dotted lines.

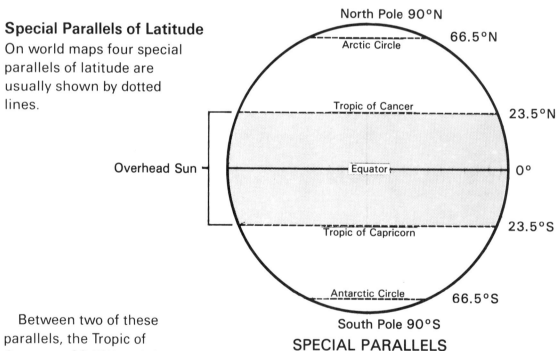

SPECIAL PARALLELS

Between two of these parallels, the Tropic of Cancer at 23.5°N and the Tropic of Capricorn at 23.5°S, lies the area that at some time during the year has the sun directly overhead. The sun will never be directly overhead north of 23.5°N or south of 23.5°S.

The two other special parallels are the Arctic Circle at 66.5°N and the Antarctic Circle at 66.5°S. Between the Arctic Circle and the North Pole lies an area in which the sun does not set during the summer. At the North Pole this daylight time lasts about 6 months. The sun does not rise in this area during the winter. This gives the North Pole about 6 months of darkness during the winter. A similar pattern occurs between the Antarctic Circle and the South Pole during its summer and winter seasons.

ACTIVITY 22
Use the maps on pp. 54-55 of the atlas to answer the following.
1. Through which large Canadian island does the Arctic Circle pass?
2. Through which North American country does the Tropic of Cancer pass?
3. Name the South American countries through which the Tropic of Capricorn passes.
4. Which ocean is located north of 66.5°N?
5. Refer to the map on atlas pp. 94-95 to answer the following:
   (a) Use the map scale to measure the distance from the Tropic of Cancer to the southern tip of India.
   (b) Estimate the percent of India that would have the sun directly overhead at some time during the year.
   (c) What effect would the overhead sun likely have on the temperature of most of India?
6. On pp. 140-141 of the atlas find which continents are cut through by the Tropic of Cancer; the Tropic of Capricorn.
7. Which continents extend north of the Arctic Circle?
8. Name a continent that would never have the sun directly overhead at any time during the year.

High, Mid, and Low Latitudes

You will often hear references to high, mid, and low latitudes. For convenience, the distance from the equator to the North or South Poles is divided into three regions: 0° to 30°N or S are the low latitudes; 30°N or S to 60°N or S are the mid latitudes, and 60°N or S to the poles are the high latitudes.

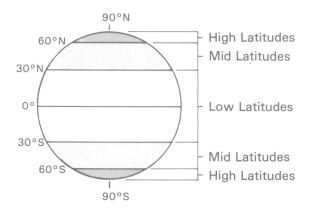

ACTIVITY 23
1. Classify each of the following as high, mid, or low latitude:
   (a) Europe
   (b) South America
   (c) Antarctica
   (d) United States
   (e) Japan
   (f) Your home
2. Calculate the width in kilometres of the low latitude region (from 30°S to 30°N).
3. Suggest advantages of living in each of:
   (a) High Latitudes
   (b) Mid Latitudes
   (c) Low Latitudes
4. Refer to the natural hazards map on p. 141 of the atlas. Which natural hazards are associated most with the low latitudes?
5. Refer to the maps of earthquakes and volcanoes on atlas p. 141. Is there any association between high, mid, or low latitude and the frequency of earthquakes and volcanoes? Why?

Meridians of Longitude

The north-south lines making up the earth grid are called meridians (from the Latin meaning, "mid-day") of longitude. The meridians extend from the North Pole to the South Pole. Any one of them could have been used as a beginning line for measuring longitude. However, to avoid confusion, it was necessary to agree upon one as a starting line. The meridian running through an observatory at Greenwich, England, was agreed upon at an international meeting held in 1884.

This 0° meridian is known as the Greenwich Meridian or the Prime Meridian. Going west or east from Greenwich, the meridian numbers increase until a point exactly halfway around the world is reached. The meridian at this point is 180°.

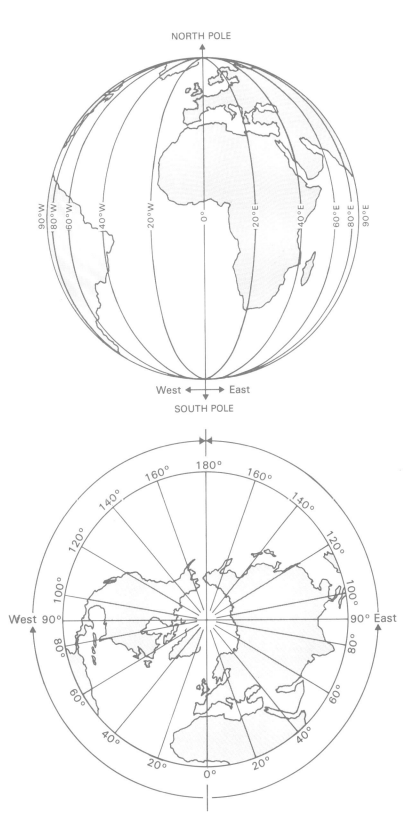

This system of meridians can be used to divide the earth into hemispheres known as the Western Hemisphere and the Eastern Hemisphere.

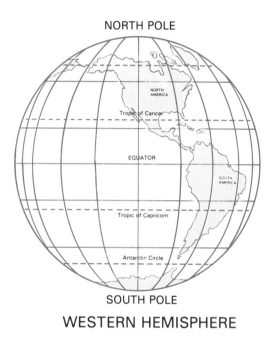

WESTERN HEMISPHERE   EASTERN HEMISPHERE

Unlike latitude lines, meridians get closer together as they move away from the Equator. At the Equator the distance between two meridians of longitude is approximately 111 km. At the poles it is 0 km.

Meridians of longitude are measured by the angle they make with the Prime Meridian at the centre of the earth.

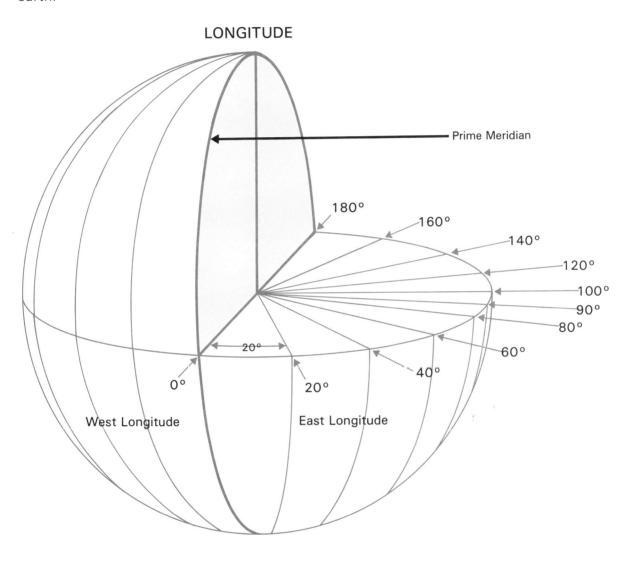

LONGITUDE

Meridians are numbered from the Prime Meridian at 0° to 180° going east and to 180° going west. Since the 180° meridian is the greatest number of degrees reached by going east or west it is not followed by a letter to indicate direction. A map will not show all meridians because it would be too crowded. How many degrees apart are they drawn on the atlas map on p. 54? How far apart are they on the maps on pp. 40-41 and 140-141 of the atlas?

ACTIVITY 24

1. Which ocean is located completely within the Eastern Hemisphere? (pp. 140-141 of the atlas)
2. Name a continent located in all four hemispheres.
3. Name the three hemispheres in which Antarctica is located.
4. Within which two hemispheres is North America located?
5. Copy the diagram and complete the numbering of the meridians of longitude.
6. Give the longitude of A, B, and C.
7. Place a small circle at the following locations: 32°E; 4°E; 25°W.
8. Refer to the maps on pp. 32-36 of the atlas and give the longitude of the following:
   (a) Winnipeg, Manitoba
   (b) Saskatoon, Saskatchewan
   (c) Calgary, Alberta
   (d) Kenora, Ontario

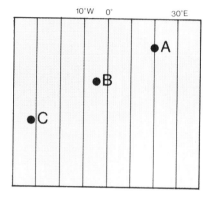

The Earth Grid

When a ship at sea is in trouble, the radio operator will give its latitude and longitude so that other ships will know exactly where to find it. You can also locate places easily on a map if you know their latitude and longitude.

ACTIVITY 25

1. Copy the diagram and label the Equator and the Prime Meridian.
2. Label the other parallels and meridians.
3. Give the latitude and longitude of each of A, B, C, and D.
4. Place a small circle at each of the following locations:

	Latitude	Longitude
E	22°S	32°W
F	20°S	0°
G	0°	20°W

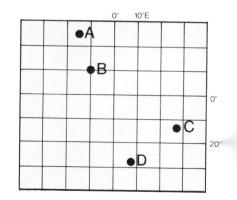

5. Give the latitude and longitude of the following:

	Atlas Page No.	Place
(a)	36	Ottawa
(b)	74	Mexico City
(c)	89	Rome, Italy
(d)	102	Paris, France
(e)	110	Bombay, India
(f)	114	Tokyo, Japan
(g)	129	Cairo, Egypt
(h)	—	Your Home

6. Find what is located at the following:

	Atlas Page No.	Latitude	Longitude
(a)	71	42°N	71°W
(b)	88	52°N	21°E
(c)	102	56°N	37°E
(d)	103	40°N	116°E
(e)	121	41°S	175°E
(f)	76	35°S	56°W

Refer to the Mercator map on atlas pp. 136-137.

7. How many degrees apart are:
   (a) the parallels of latitude shown on the map;
   (b) the meridians shown on the map?
8. On which meridian is the map centred?
9. At approximately what latitudes (north and south) has the map been cut off?
10. Give the latitude and longitude of the following:
    (a) Vancouver, B.C.
    (b) Nuuk, Greenland
    (c) Winnipeg, Manitoba
    (d) Denver, Colorado
    (e) Recife, Brazil
    (f) London, England
    (g) Canberra, Australia
11. Find the places located at each of the following:

	Latitude	Longitude
(a)	46°N	74°W
(b)	39°N	77°W
(c)	23°S	43°W
(d)	7°S	40°E
(e)	41°N	29°E
(f)	26°S	28°E
(g)	6°S	107°E

Since parallels of latitude run east-west, and meridians of longitude run north-south, both sets of lines forming the earth grid can also be used to find direction.

12. Copy the diagram and draw an arrow from A pointing north.

13. Which arrow points north?

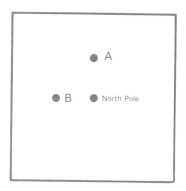

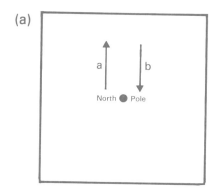

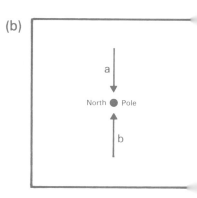

14. The arrow from Y to X points in which direction?

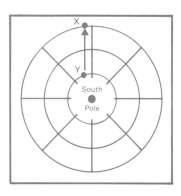

15. Copy the diagram. Begin at G and draw an arrow pointing north.

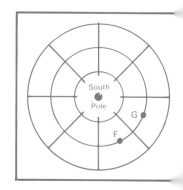

16. What direction is F from G?

17. Which point on this sketch is farther north?

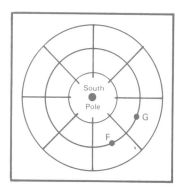

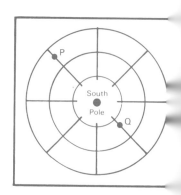

18. Which point on this sketch is farther north?

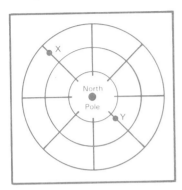

Refer to the map on pp. 20-21 of the atlas to answer the following.

19. In which direction would you travel from Fort Providence (61°N 118°W) to reach Eskimo Point (61°N 94°W)?
20. In which direction would you travel from Coppermine (67°N 115°W) to reach Churchill (59°N 94°W)?
21. In which direction would you travel to reach Aklavik (68°N 135°W) from Fort Good Hope (66°N 128°W)?

Refer to the map of Antarctica on p. 160 of the atlas to answer the following.

22. Which way is north on this map?
23. In which direction would you travel from Vostok base to reach Casey base?
24. In which direction would you travel from Byrd Station to reach General Belgrano base?

Refer to the map of the Arctic (p. 1 of the atlas) to answer the following.

25. Which way is north on this map?
26. In which direction would you travel from the North Magnetic Pole to reach the Greenland Sea?
27. In which direction would you travel from the North Magnetic Pole to reach Severnaya Zemlya?
28. In which direction would you travel to reach Nordvik from Kuujjuaq?
29. In which direction would you travel to go from Aklavik to Verkhoyansk? Nain to Arkhangel'sk?

45

GAZETTEER

One easy way to find the latitude and longitude of a place is to look it up in the gazetteer at the back of the atlas.

A gazetteer is an alphabetical listing of place names. In some ways it is like an index in a book. Beside the name of each place in the gazetteer is shown what it is (e.g. city, river), the page number of the map in the atlas where you can find the place, and its latitude and longitude. Page numbers show the largest scale map on which the feature appears, and locations are given to the nearest degree of latitude and longitude.

e.g. Port Elizabeth: South Africa **130** 34S 26E

This atlas has two gazetteers. The Gazetteer of Canada (pp. 2B-8B) lists all the names shown on the maps of Canada. The World Gazetteer (pp. 9B-24B) lists only the more important places and features shown on the world maps. Some names not listed in the gazetteer may be found on a map.

At the front of the gazetteer is a list (p. 1B) to explain the meaning of abbreviations.

Sample Listing (p. 4B)

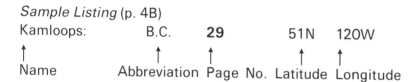

Kamloops: B.C.   **29**   51N   120W
Name   Abbreviation   Page No.   Latitude   Longitude

ACTIVITY 26

Refer to the letter U section of the Canada Gazetteer (atlas p. 7B) to complete the following.
1. How many lakes are listed?
2. How many provinces or territories are represented in the list?
3. On which page would you look for Uxbridge?
4. Give the latitude and longitude of Uivak, United States Range, Utikuma Lake.
5. What do the letters following each latitude and longitude position indicate about the hemispheres in which all of these places are located?

Refer to the letter Y section on atlas p. 22B in the World Gazetteer to complete the following.
6. On which page would you find Yeysk, U.S.S.R.?
7. In which country is Yamuna located?
8. What place is located closest to the Prime Meridian?
9. How many of these places are located south of the equator?
10. Which one of these places is located farthest east? farthest west?
11. Which two places are closest to the equator?
12. Suggest why no latitude and longitude listings are given for Yugoslavia.
13. Look at p. 1B of the atlas to find the meaning of the abbreviations cap. and Reg.

47

TIME ZONES

The earth makes one complete turn from west to east every twenty-four hours. It is daylight on the part of the earth that is turned towards the sun, and night on the part of the earth away from the sun.

No matter where you live, the sun reaches its highest point in the sky at noon. It is then noon all along a meridian of longitude from the north to the south poles. In the diagram, noon occurs at the same time in Ottawa; Philadelphia; Barranquilla, Colombia; and Huancayo, Peru on the 75°W meridian.

The earth turns through 360° of longitude (180° of west longitude and 180° of east longitude) in twenty-four hours. In one hour it turns 360° ÷ 24, or 15°. As the earth turns from west to east, the noon position of the sun reaches points farther west on the earth. When it is noon at the 75°W meridian it is still only 11:00 a.m. at 90°W longitude.

Similarly, when it is noon on the 75°W meridian, it is already 1:00 p.m. at 60°W longitude.

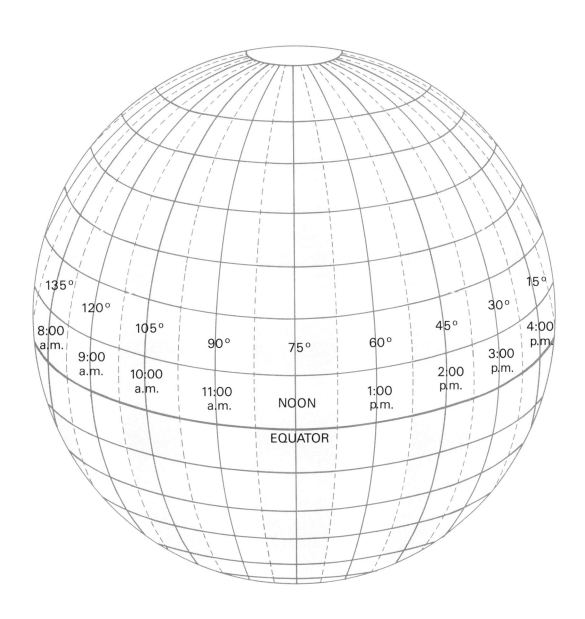

The earth is turning all the time and the noon position of the sun moves steadily across its face. It would be confusing if every place had a different time, so the earth has been divided into twenty-four time zones each 15° across. Every place within that time zone has the same time. For example, the Prime Meridian is the centre of a time zone that is 15° wide.

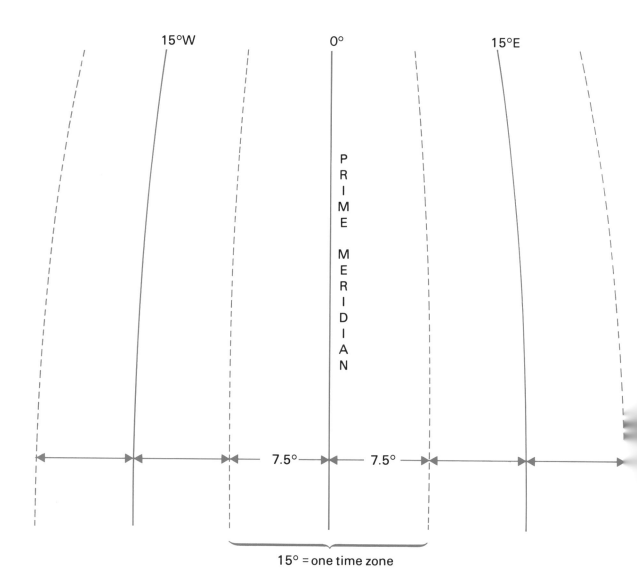

When it is noon at the Prime Meridian, halfway around the globe at 180° it will be midnight. Therefore, between the 0° and the 180° meridian there will be 12 time zones going west and 12 time zones going east.

ACTIVITY 27

Refer to the inset map of world time zones on p. 138 of the atlas to answer the following.

1. Why do time zone boundaries not always follow straight meridians?
2. Which country in the world spans the greatest number of time zones? How many are there?
3. How many time zones are there in the United States?
4. What is your local time if you are 105°E of the Prime Meridian and Greenwich Mean Time (G.M.T.) is 6:00 a.m.?
5. You are a ship's captain who has calculated the latitude of the ship as 20°N. Your local time is 10:00 a.m. when G.M.T. is 4:00 p.m. In which body of water are you located?
6. Examine the three clock faces. Each shows the same point in time but each is at a different meridian. Which clock is at a meridian between the other two clocks? If G.M.T. is 12:00 noon, on which lines of longitude are each of the clocks located?

The International Date Line

Starting with the Greenwich meridian as 0° and going eastward, there are 12 time zones of 1 hour each to the 180° meridian. There are another 12 time zones of 1 hour each going westward to the 180° meridian. These total 24 hours or one full day. This makes the 180° meridian a natural place to change dates. Since very few people live along this meridian it causes little inconvenience to make this change. Because of its importance as the line along which one day disappears and the next begins, this meridian was named the International Date Line at the International Meridian Conference held in Washington, D.C. in 1884.

As you can see from the diagram on the right, the calendar day on the Asiatic side of the 180th meridian is later than the calendar day on the American side. Thus, if it is Monday on the Asiatic side, it is Sunday on the American side.

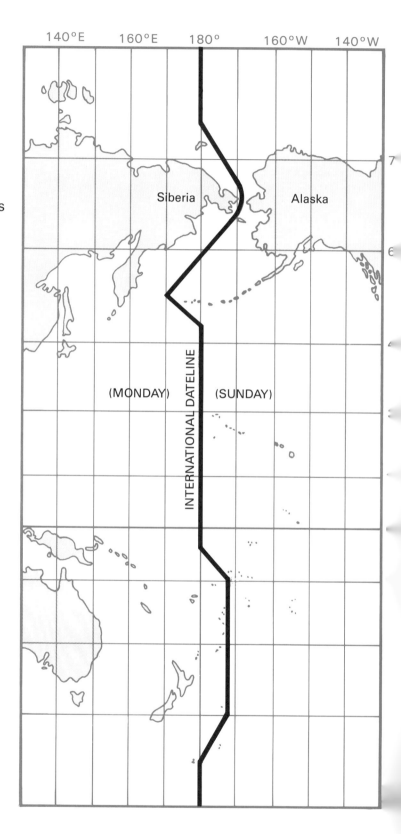

Time zone boundary. This sign is located on the boundary between the Mountain and Pacific time zones. When you travel west from Jasper National Park to Mount Robson Provincial Park a one hour change in time takes place here.

Time zone boundary. When travelling east from Mount Robson Provincial Park to Jasper National Park, a one hour change in time occurs here.

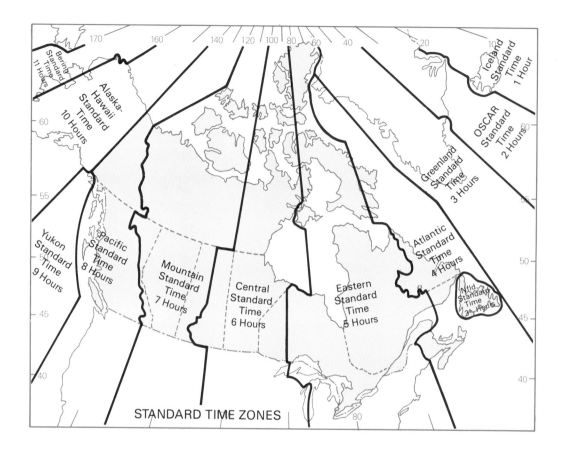

ACTIVITY 28

Use the time zone map of Canada to answer the following.

1. How many time zones are there in Canada?
2. Why do the boundaries of time zones not always follow a straight line?
3. Name your home time zone, the standard meridian on which it is based, and the number of hours time difference between your zone and the Greenwich zone.
4. If a hockey game is televised at 8:00 p.m. in Toronto, what time will it be seen in Vancouver?
5. How big a time difference is there between the Newfoundland and Atlantic time zones?

SCALE

It would be difficult to draw a map of your classroom as large as the classroom itself. It would be impossible to draw a map of the world as large as the world.

The drawing to the right represents a peanut butter and jam sandwich.

In this diagram the sandwich is about 3 cm wide. A real sandwich is about 12 cm wide. This drawing is a "map" of a real sandwich but four times smaller.

The drawings below show the same sandwich drawn two other ways.

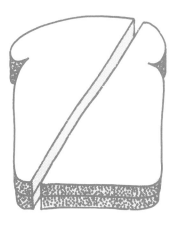

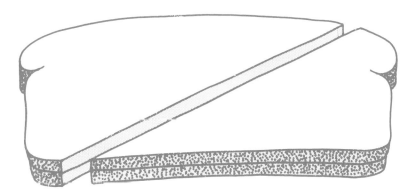

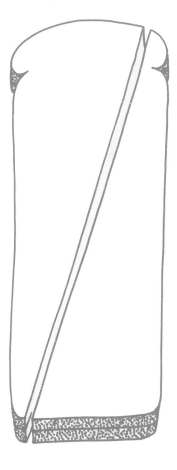

ACTIVITY 29

1. In what ways are each of these different in shape from a real sandwich?
2. What would have to be done to each of these to make them look like a real sandwich?
3. Why could these not be considered "maps" of a real sandwich?

Map Scale

A map represents the earth's surface or part of the earth's surface drawn smaller than it really is. A map is drawn to scale. Scale means that a small distance on the map stands for a much larger distance on the earth's surface. The scale that is printed on a map shows us the actual distance between places on the earth's surface.

Most maps have a scale printed on them. There are three ways to show scale.

1. *Statement scale*

    1 cm : 50 km

This means that 1 cm on the map stands for 50 km on the earth's surface. To find a distance on the ground you measure the distance between two points on the map in centimetres. Then calculate the actual distance in km.

A ●————————— 8cm —————————● B

The map distance between A and B is 8 cm.
The real distance would be 8 × 50 = 400 km.

2. *Line scale*

In a way the line scale is like a special kind of ruler to use with the map. Each space of 1 cm on the line scale stands for 50 km on the earth's surface.

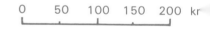

Turn to the map on p. 32 in the atlas. Place a piece of paper with a straight edge on the map so it touches both Calgary and Edmonton. Mark the locations on your paper.

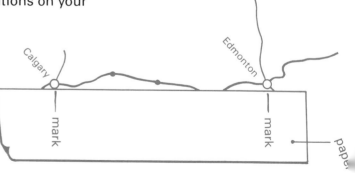

56

Now place the edge of the paper on the line scale with the first mark at 0 and mark the distance at the end of the 200 km on the paper. Move that mark to 0 and calculate the remaining distance.

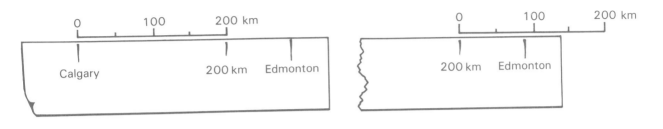

By this method you calculate the distance from Calgary to Edmonton as approximately 290 km.

3. *Representative Fraction*

$$1:5\,000\,000 \quad \text{or} \quad \frac{1}{5\,000\,000}$$

This is known as the representative fraction or R.F. scale. This scale shows that the distance on the ground is 5 000 000 times greater than the same distance on the map.

Both the statement and line scales are useful when working with metric measurements. Many maps, however, may use feet or yards or miles to measure distances. On this scale, 1:5 000 000 can mean that 1 cm represents 5 000 000 cm, or 1 inch represents 5 000 000 inches. No matter what units are used the R.F. scale helps you find the exact distance. This is why the R.F. scale is so useful. It gives a scale that can be understood by anybody in the world.

ACTIVITY 30

1. How tall is the real bottle?

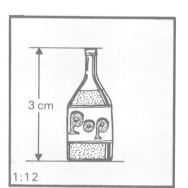

2. How long is the real bird? (Measure its length then calculate the real size.)

3. The mast on this model boat is about 4 cm long. How high is the actual mast?

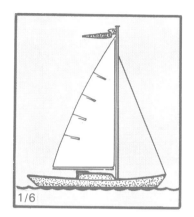

4. On this map, A is approximately 2 cm from B. How far apart are they on the surface of the earth?

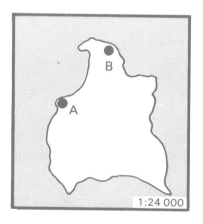

5. How far apart are places A and B? (Measure the distance between A and B then calculate the actual distance.)

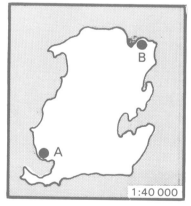

Maps drawn at various scales show different amounts of detail.

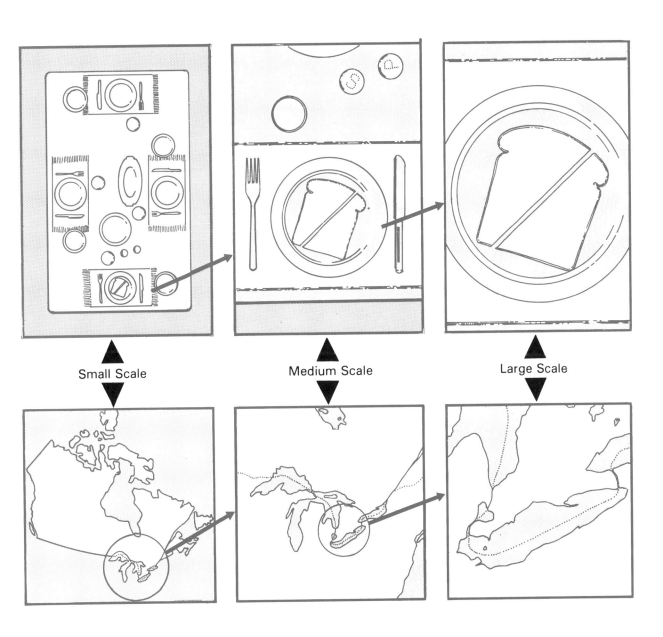

ACTIVITY 31

1. How does the amount of detail change as you move from a small-scale map to a large-scale map?
2. Which type of map would you choose if you wanted detailed information about a small area?
3. Which of the three scales is used on most atlas maps?
4. Find two large-scale maps in the atlas. What do they show?
5. Use the gazetteer to locate Algiers, Algeria and Khartoum, Sudan. Find them on the maps on pp. 128-129 of the atlas and use the map scale to calculate the shortest distance between them.
6. Locate Bogota, Colombia and Rio de Janeiro, Brazil on the map on p. 76 of the atlas. Use the map scale to calculate the shortest distance between them.

Scales of Atlas Maps

The page size of the atlas is approximately 22.5 cm wide and 29 cm long. Although the area of the page cannot vary, different parts of the world of different sizes are shown on the map pages — all the way from a part of Canada, or province, to a whole continent, or even the whole world.

The scales used on the atlas maps vary depending on the size of the area to be shown on the atlas page.

ACTIVITY 32

1. Below is part of a chart. Copy the headings and first line, which has been done for you, into your notebook.

REFERENCE	MAP TITLE	ATLAS SCALE	1 cm:	5 mm:	1 mm:
Atlas page A p. 10	Canada: Population	2.7 cm : 1000 km	370 km	185 km	37 km

Complete the chart for the following:
B pp. 2-3
C pp. 22-23
D pp. 28-29 (the large map)
E p. 28 (the small map)
F p. 27

2. The lines below represent distances on maps. The letter following the line refers to the scale on the chart you have compiled. First, measure the length of each line in centimetres and millimetres. Then use the scale to calculate the actual distance in kilometres.

1. ———————————————————— F
2. ————————— A
3. ————————————— C
4. ——— E
5. ——————————— B
6. ————————— D
7. ———————————————————— A
8. ————————— F
9. ————————— E
10. ————————— F

ACTIVITY 33

The outlines of the British Isles are drawn below at a variety of scales.

1. What happens to the size of the 1 cm line and the 1 cm² as the map scale increases?
2. What happens to the distance represented by the 1 cm line as the map scale increases?
3. What happens to the area of land represented by 1 cm² as the map scale increases?

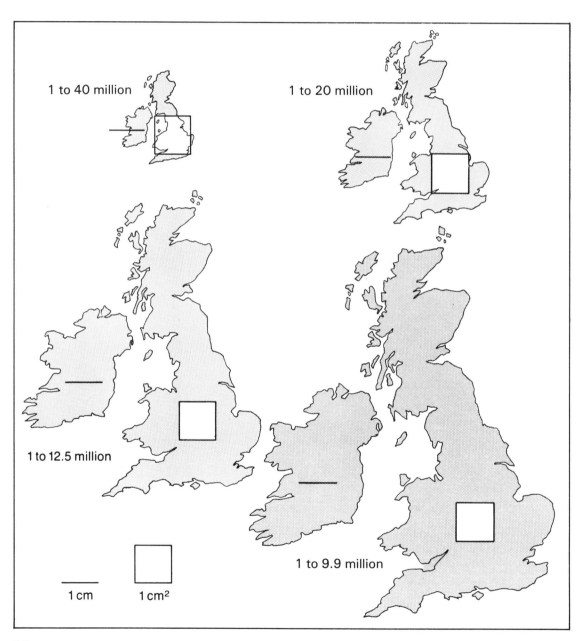

4. Suggest types of information that might be shown on:
   (a) a map at a scale of 1 to 40 million;
   (b) a map at a scale of 1 to 6.6 million.
5. Refer to the following atlas maps: pp. 2-3; physiographic regions, p. 8; pp. 20-21; your province or territory, pp. 20-21, 28-43. Find the following for each map:
   (a) What is the map scale?
   (b) What distance is represented by a 1 cm line?
   (c) What area of land is represented by 1 cm^2?
   (d) What happens to the distance represented by the 1 cm line as the map scale increases?
   (e) What happens to the area represented by 1 cm^2 as the map scale increases?
   (f) What types of information are shown on the largest and smallest scale maps in this question?

ACTIVITY 34

Refer to the maps of China, atlas pp. 110-111 and 112-113.

1. Draw and complete this chart:

MAP		REPRESENTATIVE FRACTION	LINEAR SCALE
(a)	pp. 110-111		8 mm:
(b)	pp. 112-113		8 mm:

2. (a) Which of these two maps shows the most detail? the least detail?
   (b) Which one is most useful for making comparisons with neighbouring countries?
3. Write the two representative fraction scales in your notebook. As the size of the ratio in each representative fraction increases, what happens to the size of the land area that can be shown on a map? the amount of detail that can be shown?
4. On both of these maps, the linear scale is divided into units of 8 mm. Against each representative fraction in your notebook, write the number of kilometres represented by the 8 mm segment on that map.
   (a) What happens to the distance represented by each unit as the size of the ratio in the representative fraction decreases in value? How does this affect the amount of detail that can be shown?
   (b) On atlas pp. 110-111 the page area covered by the map is approximately 37 cm × 25 cm. Use the linear scale to calculate the actual area shown by the map. Make the same calculation for pp. 112-113, using the linear scale for that map. How does the size of the land area shown on each map relate to the size of the ratio in the representative fraction scale?

5. Compare the apparent size of Japan on p. 111 and on p. 114 of the atlas. What is the representative fraction scale on each of these two maps?
6. Compare atlas pp. 110-111, 112-113, and p. 114 to the world physical map on atlas pp. 140-141:
   (a) What is the representative fraction scale on the world map? How does it compare to the others?
   (b) What happens to the amount of detail and area coverage on the world map when compared to the other maps?
7. Why is it necessary to check the map scales before comparing the sizes of countries on different maps?

TYPES OF MAPS USED IN THE CANADIAN OXFORD SCHOOL ATLAS, 5TH EDITION

The atlas contains several maps of each part of the world. Each map presents different kinds of information about that part of the world.

The types of maps used in this atlas are:
place name or political maps
thematic maps
physical maps
regional topographic maps

PLACE NAME OR POLITICAL MAPS

The map of the world on pages 136-137 of the atlas shows all the countries of the world. There are also maps of each continent (Europe, etc.) that show all the countries and major cities of that continent. The map of Canada on atlas pp. 2-3 shows the provinces and major cities in Canada. These maps are useful because they give a general picture of the location of countries and cities within a large area.

The legends on these maps do not give information about the colours used on the maps, because the colours are used only to distinguish each country or province from its neighbour.

The legends give other information appropriate for the particular map. For example, the political map of the world gives only the names of the countries and the capital city of each country, so no legend for the place names is needed. Other political maps give information on the cities shown on them: see the legends on atlas pp. 65 and 93. Even more detail is shown on the map of Canada; the legend on p. 3 gives symbols for many different sizes of settlements.

In addition to the legend, different sizes and different styles of type are used on maps to show different things.

ACTIVITY 35
Look at the map of Canada on atlas pp. 2-3.
1. How are the names of the countries shown?
2. How are the names of the provinces of Canada shown?
3. How is the name of the capital city of each province shown?
4. Look at Vancouver, Prince George, and Penticton. What is the difference in type for each of these three settlements?

ACTIVITY 36
1. Examine the Table of Contents and list the titles of all the political maps in the atlas.
2. Refer to the map of Asia on atlas p. 93.
   (a) List all the countries that share a border with each of the following:
      (i) India
      (ii) China
      (iii) U.S.S.R.
   (b) Name the country in (a) that has the most neighbours.
   (c) Which of India, China, U.S.S.R. share a border?
   (d) Name a country that shares a border with all three of the countries in (a).
3. Refer to atlas pp. 30A-32A.
   (a) Use information from the tables to compare the following for India, China, and the U.S.S.R.:
      (i) Area
      (ii) Population
      (iii) Population Density
      (iv) Arable and Permanent Pasture (farmland)
      (v) Urban Population
   (b) Rank order the countries for each of the categories listed above.
   (c) Name the country that ranks second in three of the five categories.
   (d) Name the country that has the largest percentage of its population classified as rural.

THEMATIC MAPS

Thematic maps give specialized information about the world, the continents, and Canada. They use a variety of colours and symbols to present information on such themes as people and their activities, the climate, and the natural environment. Two examples of thematic maps are shown on atlas pages 100-101. On each map a range of colours is used to show different information. Each colour on each map represents areas that have the same characteristics.

ACTIVITY 37
Refer to the maps on atlas pp. 8-9.
1. (a) What four themes are shown on these maps?
   (b) One colour on the upper map on p. 8 is used to show "Cold winter and hot summer. Very dry in the south." The same colour is used on the lower map. What does it show there? The same colour is used on a map on atlas p. 17. What does it show there?
2. Examine the table of contents of the atlas and list the variety of themes used in the atlas.
3. (a) Refer to the gazetteer to locate the following cities:
       Tokyo
       Paris
       New Delhi
       Edmonton
   (b) Examine the appropriate thematic maps to note the following for each of these cities:
       Country   Soil
       Latitude   Life Expectancy
       Longitude   Religion
       Climate region   Nutrition
       Vegetation

PHYSICAL MAPS

The physical map of the world on atlas pp. 140-141 names and shows the major physical features of the earth: the mountain chains, the major rivers, lakes, etc. Physical maps of Canada and of each continent show the physical features in more detail.

The surface of the earth is not flat. Physical maps use different colours to show the varying heights of the land — mountains, plains, and valleys — and the varying depths of water in the oceans.

ACTIVITY 38

Turn to pp. 94-95 of the atlas.
1. List the different types of physical features shown on this map under the headings Land Features and Water Features.
2. Name the mountain range that forms the boundary between Europe and Asia.
3. (a) What is represented by the symbol on the right?
   (b) Give the page number of another map in the atlas that has this same symbol on it.
4. (a) What is the name of the area of highest elevation in Asia?
   (b) What is the altitude of Mount Everest?
5. Name the water body that has an area of land depression around it.
6. Name the body of water that has the greatest water depths.
7. Name two major rivers that flow into each of:
   Arctic Ocean   Indian Ocean
   Pacific Ocean   Black Sea

REGIONAL TOPOGRAPHIC MAPS

There are many maps of this type in the atlas. These maps are at a larger scale than most other maps in the atlas, and therefore show a smaller area, or region, in more detail. For example, there are regional topographic maps of each province or group of provinces in Canada; and of regions of each continent, such as Eastern U.S., Western Europe, and Southeast Asia.

Because these maps are drawn at a larger scale, they can show many more places than the political maps; they also show other cultural features such as major roads and railways.

The maps are called topographic because, in addition to the features described above, they use colours to show the height of the land, and they name the major physical features. From these maps it is possible to study the interrelationship of physical and cultural features.

These maps are also the most useful maps for finding the location of cities and towns; most references in the gazetteer are to these maps.

Regional topographic maps have a key to the colours used on the map (land heights), and a key to the lines and symbols used on the map. For all the topographic maps in this atlas other than those of Canada, the legend for the size of settlements is shown in the shaded area on atlas p. 1.

ACTIVITY 39
Refer to the map on atlas pp. 106-107.
1. (a) What is the difference in the representative fraction scale between this map and that on atlas pp. 94-95?
   (b) What is the difference in the distance represented by 1 cm on each of these maps?
   (c) Use the linear scales to calculate the approximate areas of each of these maps.
2. What information is included in the legend on this map that is not included in the legend of the map on atlas pp. 94-95? Why?

3. Name the two major rivers that empty into The Gulf.
4. (a) Refer to the gazetteer to find the latitude and longitude of the following and then locate each one on the map on pp. 106-107 of the atlas: Urfa, Arak, Zabul, Baku, Dammam, Ismailia, Buraydah.
   (b) For each of the places above name:
      (i) one in a land depression;
      (ii) one in a marsh;
      (iii) two on oil pipelines;
      (iv) two at an elevation of 500-1000 m;
      (v) one at an elevation of 2000-3000 m;
      (vi) one on a canal;
      (vii) one in a sand desert.

MAP PROJECTIONS

The only accurate map of the earth is a globe. Provided it is large enough, all parts of the earth's surface can be represented on it in their true shape, relative size, and position.

However, most of the maps you use are printed on flat sheets of paper. Flat maps can be easily rolled, folded, bound into atlases, and stored. They are also easy to carry around. But a flat map of the earth can never be truly accurate. It is impossible to flatten a global surface without it stretching, wrinkling, or tearing.

ACTIVITY 40

Imagine that you have a piece of a burst balloon, pointed at one end, and with a face painted on it, like this:

1. Now suppose you stretch out the strip of balloon so that it fills the square outlined here by the dotted line:

What would happen to the shape of the eyes?

2. Do you think the mouth would stretch as much as the eyes would?

3. (a) Now suppose you start with a piece of map instead of a balloon. Where would the map have to be stretched the most to fit the rectangle?

   (b) Which parts of the world would stretch out of shape the most?

All flat maps distort the part of the earth they represent. When a map represents only a small part of the earth's surface, such as a city or province, the distortion is not great. However, when it shows a continent, hemisphere, or the whole world the distortion is much greater.

One way to make a flat map from a globe could be to cut the globe along the meridians. This separates it into a number of sections (usually 12 or 24). The map that results is difficult to use. Some of the land masses, e.g. Europe, are divided in two. Others, e.g. Asia, are cut into many sections.

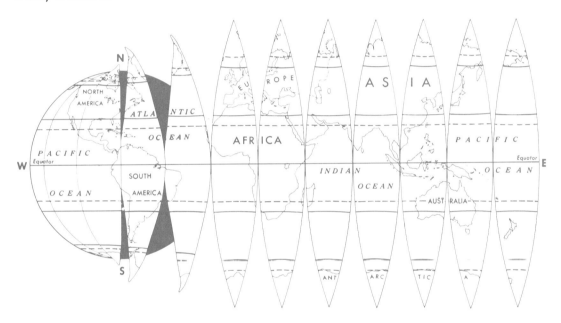

ACTIVITY 41

1. Here is the whole world in strips. Which of the following is correct?
   (a) there is a little bit of Antarctica in each bottom point;
   (b) all of Antarctica is in one of the bottom points.

2. What do you think would happen to Antarctica if you stretched each segment until the map filled the rectangle?

3. Do you think the areas near the Equator would stretch as much as those near the Poles?

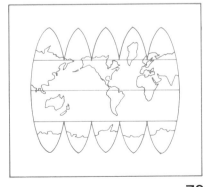

To make a map that is easier to use, a cartographer (map-maker) rearranges the globe grid on a flat surface in such a way that one or more of the globe's geometric properties can be kept. These properties are direction, distance, shape, and area. The cartographer keeps the worst distortion to the least important area of the map. For example, some of Europe might be included on a map of Asia. Because it is not the focus of the map and appears only at its edge, it is more likely to be distorted.

It is possible to construct a framework of parallels and meridians so that land areas plotted on them show true area, but only by sacrificing their shape. On the other hand, true shape can be kept but only at the expense of area.

A map projection is a way of projecting the globe grid of parallels and meridians onto a map. Very few projections are actually "projected" in the true sense of the word.

No one map projection is better than all others. Each one has certain qualities that make it useful for a specific purpose. And since there are many ways to overcome distortion, there are hundreds of different map projections. It is likely, however, that most maps you use will be drawn on one of the ten or twelve most commonly used projections.

A MAN'S HEAD PLOTTED ON GLOBULAR, STEREOGRAPHIC, AND MERCATOR PROJECTIONS

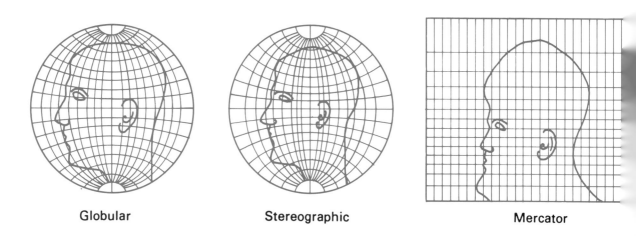

Globular   Stereographic   Mercator

Deetz & Adams, *Elements of Map Projection*, U.S. Coast and Geodetic Survey. Special Publication #

ACTIVITY 42
1. (a) Which one of the projections on the previous page has the greatest distortions?
   (b) In which parts of the projection do the greatest distortions occur?

2. Refer to the map below (Mercator projection).
   (a) Which of A or B represents the shorter distance? Verify your answer by using a piece of string on a globe.
   (b) Use a globe and a piece of string to find the shortest routes between:
   Toronto and Moscow
   Vancouver and Tokyo
   Halifax and Peking

   Mark these on a wall map and note the major geographical areas they cross.

3. What are some of the disadvantages of a globe as a convenient reference?

Map Projections used in the Canadian Oxford School Atlas, 5th Edition

The following six projections are used for all the maps in the atlas:
   Conical Orthomorphic
   Oblique Mercator
   Transverse Mercator
   Modified Gall
   Zenithal Equidistant
   Zenithal Equal Area

In some cases the word "Modified" is used in front of projections other than Gall's to show that the cartographer has changed the original projection to increase the accuracy of one or more of distance, area, shape, or direction.

CONICAL ORTHOMORPHIC

This is the most common projection used in the atlas. The word "orthomorphic" tells us that the shapes of small areas are reasonably well preserved.

A cone placed over a globe cuts it at two standard parallels — in this case 30°N and 60°N.

When the cone is unrolled a portion of the grid would have properties like this.

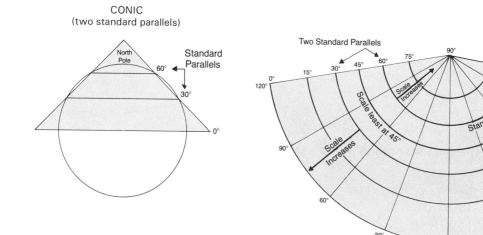

The region between the two standard parallels is reasonably accurate in shape and area, as are the narrow strips beyond the standard parallels. The important decision for the cartographer is to make the right choice of standard parallels for the area to be mapped.

ACTIVITY 43

Refer to maps on atlas pp. 20-21, 104-107, and 110-111.
1. Where would the meridians meet if they extended beyond the maps?
2. (a) Do most of these maps have greater latitudinal or longitudinal extent?
   (b) Why is the greater extent suited to this projection?
3. What size areas are shown best by this projection?

CYLINDRICAL

Several projections in the atlas are based on a cylinder rather than a cone. Cylindrical projections are based on the idea of a cylinder surrounding the globe and touching it along the equator. When the cylinder is unrolled it forms a rectangle with a length equal to that of the equator.

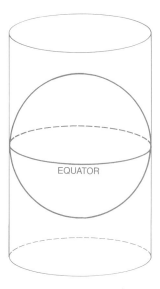

1. Mercator

Gerardus Mercator was a Dutch geographer who devised a projection in 1569 that helped seamen find accurate compass bearings. This projection remains the standard for navigation charts. Unfortunately it was so widely accepted that it has been used for many purposes for which it was never meant. The result has been many mistaken impressions about the relative sizes of land masses on the earth.

The Mercator projection is a mathematical adaptation of a cylindrical projection based on the equator. To keep the map conformal (to preserve shape), parallels and meridians become increasingly far apart poleward. The distortion of scale and area becomes so great that the map is usually cut off at about latitude 80° or 85° N and S. The poles can never be shown. Shapes are true for all small regions.

Uses

This projection is very suitable for air and sea navigation and for any other map where correct direction is necessary, for example those showing wind direction or ocean currents.

ACTIVITY 44
Refer to the map on atlas pp. 136-137.
1. (a) What happens to the space between the parallels as you move north and south of the equator?
    (b) What effect does this have on scale and area?
2. (a) What happens to the meridians as you move polewards?
    (b) What effect does this have on scale and area?
3. What parts of the globe cannot be shown on this type of map?

2. Oblique Mercator

If a globe touches the cylinder neither at the equator nor along a pair of meridians, it is an oblique projection.

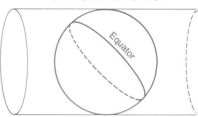
OBLIQUE MERCATOR

ACTIVITY 45
Refer to the maps on atlas pp. 54-55.
1. Which land masses are represented on these maps?
2. Which is the only parallel that is almost a straight line?
3. What is the shape of the parallels of latitude?

3. Transverse Mercator

This projection results from a cylinder touching the globe along a pair of central meridians. Areas near the central meridians combine correct shape with little error in scale. It is particularly good for areas of land with north-south dimensions.

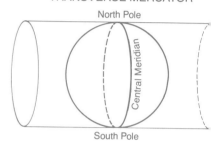
TRANSVERSE MERCATOR

ACTIVITY 46
Refer to the maps on atlas pp. 76 and 109.
1. (a) What areas are represented on each of these maps?
   (b) Why was this a good projection for these areas?
2. What is the central meridian on each of these maps?
3. (a) What shape are the parallels on the map on p. 76?
   (b) Which parallel is the only straight line?

4. Modified Gall

Gall's is a form of map projection which presumes that a cylinder cuts the globe in two places — at parallels 45°N and 45°S. This projection is neither equal area nor orthomorphic but it does present a reasonable compromise between accuracy of shape and area.

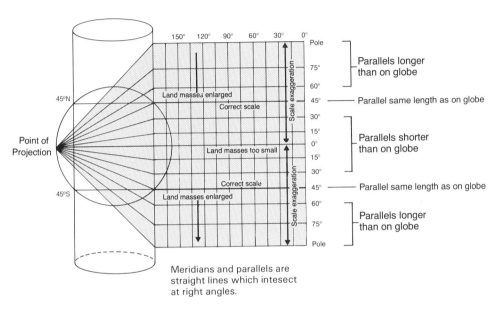

Meridians and parallels are straight lines which intesect at right angles.

With modifications, distortion of shape and area can be lessened so that this projection can be used as a reasonable general purpose map of the world, except in high latitudes.

ACTIVITY 47
Refer to the maps on atlas pp. 136-137, 145, 146-147, and 159.
1. (a) How much of the world is shown on these maps?
   (b) What is missing?
   (c) Can the missing parts ever be shown on a cylindrical projection?
2. Where does the greatest distortion occur on these maps?
3. (a) Why are the words "equatorial scale" used with this projection?
   (b) Why are these maps not useful for finding distances between places?
   (c) What projection would be better for finding distance?

ZENITHAL EQUIDISTANT

These projections are made by placing a flat sheet of paper against the globe. When the paper touches the equator the projection is known as equatorial. When the paper touches one of the poles the projection is called polar. When the paper touches any other point on the globe the projection is called oblique.

POINT OF CONTACT

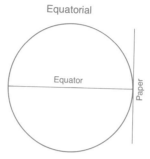
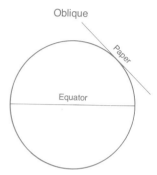

The biggest advantages of these projections is that both direction and distance from the centre of the map are correct, and that the projection can be centred on any point on the earth's surface.

ACTIVITY 48
Refer to maps on atlas pp. 1, 2-3, 114, 132-133 and 160.
1. (a) On which of these maps was the contact with the paper at the poles?
   (b) What shape are the parallels on these maps?
   (c) Where do the meridians converge?
2. (a) Which one of these maps has the equator as the point of contact?
   (b) What shape are the parallels on this map?
   (c) Is this map orthomorphic?
3. (a) How many of these maps are zenithal oblique projections?
   (b) How large an area do they represent?
4. Which one of the maps listed above shows the largest area of the world?

ZENITHAL EQUAL AREA

This projection, devised by J.H. Lambert in 1772, is constructed by a formula which gives it true equal area. It is most suitable in the oblique projection for any continent that is limited to about 40° north or south from the centre of the map. In the map of North America, area is preserved, shape is well represented, and direction from the centre of the map is correctly represented by a straight line.

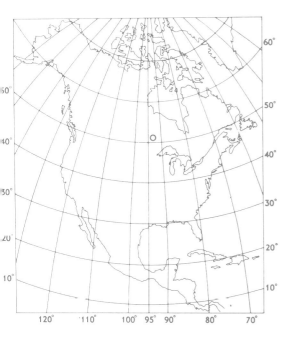

ACTIVITY 49

Refer to the maps on atlas pp. 94-95, 122, 128-129.
1. What is the most frequent size of land area shown by this projection?
2. What is the central meridian of the projection on pp. 94-95? on p. 122?
3. What shape are the parallels of latitude? the meridians of longitude?
4. What would be one of the best uses for an equal area projection map?

GREAT CIRCLE ROUTES

Suppose you are planning to take a long trip by air or ship to a distant city. The easiest way of finding your route is not on a map, but on a globe. The earth is shaped like a ball or a sphere. No flat map can give an accurate picture of any large part of this sphere.

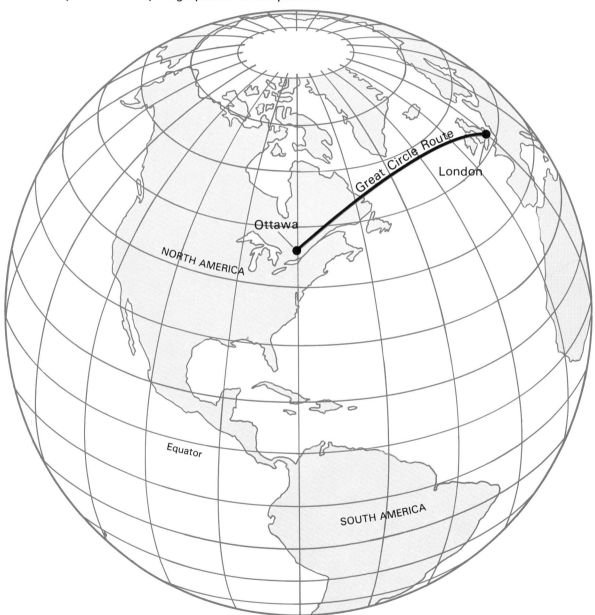

The easiest way to find the distance and route between two cities which are far apart is to use a piece of string on a globe. Suppose you want to travel from

Ottawa to London, England. When you stretch a string between them on a globe you will notice that the shortest route is not one which goes directly east. It is one that curves in a northward arc. You can check this by finding the shortest route from Vancouver to Tokyo.

These most direct routes are called great circle routes. Any circle which cuts the globe in two equal parts is a great circle. The equator and all meridians cut the globe in half and are thus great circles. The equator is the only parallel that is a great circle.

Since great circles are the shortest distances between two places on the earth's surface, airlines and ships travel in great circles whenever conditions make this possible.

ACTIVITY 50
1. Use a string to measure the following great circle distances on a globe (make sure the string doesn't stretch):
   (a) Ottawa to London, England
   (b) Vancouver to Tokyo
   (c) Recife, Brazil to Dakar, Senegal
   (d) Los Angeles to Hong Kong
   (e) Lima, Peru to Sydney, Australia
   (f) Vladivostok, U.S.S.R. to Perth, Australia
   (g) Anchorage, Alaska to Wellington, New Zealand
2. Use the map scale to measure the straight line distances between these same places on atlas maps. Refer to the following maps:
   (a) p. 131
   (b) pp. 132-133
   (c) p. 131
   (d) pp. 132-133
   (e) pp. 132-133
   (f) pp. 132-133
   (g) pp. 132-133
3. Calculate the differences between the great circle distances on the globe and the straight line distances on the maps.
   (a) Which are smaller?
   (b) What are the greatest and smallest differences? How do these relate to the distances between places?

COMPASS BEARINGS

We can use compass bearings as another way to indicate direction on maps. The directions on a diagram of compass bearings can be replaced by degrees. A circle has 360°.

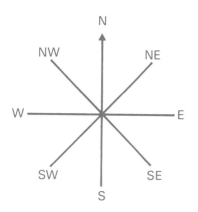
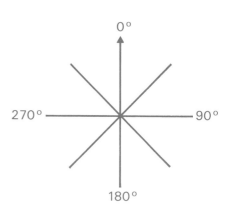

ACTIVITY 51
1. Find the bearing from A to B. First draw in a north line.

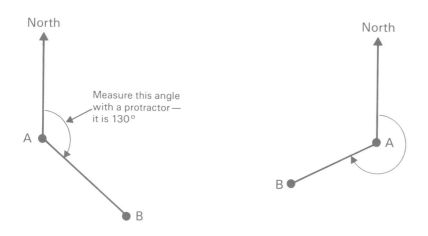

REMEMBER: When giving a bearing from A the angle from North is measured at A. Always work out bearings clockwise from North, as in the example on the right.

ACTIVITY 52
1. Copy the diagram on the previous page and fill in the missing numbers.
2. Find the compass bearings on this diagram. Go from A to B, B to C, etc.

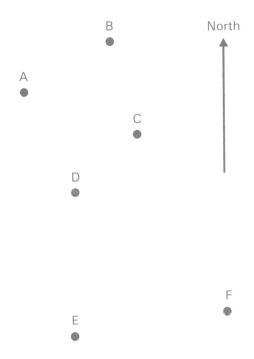

Compass Bearings on a Mercator Map

Any straight line drawn on the mercator projection map is a line of constant compass direction. The compass bearing can be readily found by joining any two places on this map with a straight line and then measuring the angle between that line and any meridian.

This line is known as a rhumb line or loxodrome. On the map on the next page, the straight line joining Buenos Aires and Cape Town crosses all meridians at the same angle, so it is a rhumb line.

However, the shortest distance between these two places lies on a great circle. This is shown by the curved line joining Buenos Aires and Cape Town. While it appears simple enough just to set out from Buenos Aires, set the compass at one heading, and go to Cape Town, most ships and planes prefer to take the shortest route, which is the great circle.

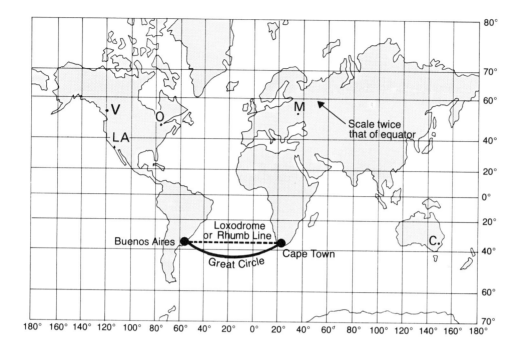

A great circle course is found by plotting a series of loxodromes along a great circle. As you can see on the diagram, the straight lines must continually break off and change direction if they are to follow the circle. The more legs there are, the closer the loxodromic route is to the great circle route.

Since the equator and all meridians are great circles, any straight line drawn along them is both a rhumb line and a great circle at the same time.

ACTIVITY 53
1. On the map above trace straight lines joining the following places and use a protractor to find the approximate compass bearings from:
   (a) Vancouver to Ottawa
   (b) Ottawa to Los Angeles
   (c) Ottawa to Canberra (across the Atlantic Ocean)
   (d) Ottawa to Moscow
2. Why is it not necessary to indicate direction after compass bearings?

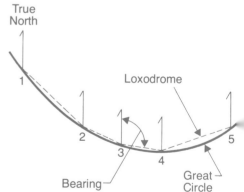

PART 3:
Applied Skills

PHOTOGRAPHS

You read books to obtain information, and you probably enjoy looking at the illustrations and photographs. But did you know that photographs are often put in books because they provide information of a different kind from the written material? You can "read" photographs as well as reading words and maps.

There are several steps to follow when reading photographs. These include:
  (a) gaining an overall impression of the scene;
  (b) identifying specific features;
  (c) finding relationships between features;
  (d) making inferences from what is seen in the photo about things that cannot be seen;
  (e) recognizing patterns.

Patterns on Photographs

There are two kinds of patterns made on the land. Natural patterns, like those made by rivers, are usually irregular in form. Patterns made by human settlement are usually much more regular in form and may involve straight lines and geometric shapes.

These are some of the more common patterns.

NATURAL PATTERNS
- Volcanic cone
- Meander

HUMAN PATTERNS
- Circular
- Rectangular
- Parallel
- Triangular
- Radial
- Oval
- Square
- Grid
- Dotted Regular
- Dotted Random

Types of Photographs

There are two main types of photographs used for studying the earth.
  (a) Ground-level photographs — these are the kinds of photographs most people take with their own cameras as they stand on the ground.
  (b) Air photographs — these are photographs taken from the air by a camera pointed at the ground.

ACTIVITY 54
1. What do you see in this ground-level photograph?
2. Use the horizon as a guide to describe the land surface. There are two general areas in a photograph.
   (a) **Foreground** — this is the part of the photograph nearest the viewer, the part that shows the most detail. It is approximately the bottom half of many photographs.
   (b) **Background** — this is the part of the photograph farthest away from the viewer, the part that shows the least amount of detail. It is approximately the top half of many photographs.
3. What pattern is displayed in the foreground of this photo?
4. What are the two dark buildings on the right side of the photo? What might they be used for?
5. What is the only visible form of transportation in the background?

6. What are the major types of land use here?
   (Land use — any use of the land such as crops, buildings, roads, etc.)
7. What season of the year is it?
8. How would the farmer get his grain to market?
9. Why is the crop left out in the field?
10. Why are there so few trees in this area?

AIR PHOTOGRAPHS

Photographs taken from the air show a great deal about the land and the patterns found on it. There are two major types of air photographs, oblique and vertical.

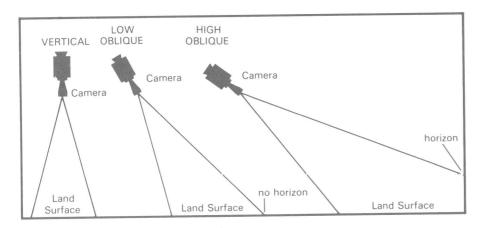

Oblique Air Photographs

Oblique photographs, such as the ones on pp. 90 and 91, are taken with the camera pointed towards the horizon. When the camera is tipped high enough to show the horizon, it is called a high oblique photograph. When no horizon is shown it is a low oblique photograph.

This type of photograph gives the kind of view seen by a passenger looking out the window of an airplane. A large amount of land can be seen, but the detail is usually limited to the foreground of the photograph. Because the camera is tilted, tall buildings in the foreground will hide objects behind them. However, the "natural" appearance of objects in the photograph makes them easy to identify.

ACTIVITY 55

Refer to the above photo of Toronto.
1. Is this a high or low oblique?
2. What do you see in this picture?
3. Use the horizon as a guide to describe the land surface.
4. Describe any three patterns in what you see.
5. Name any five land uses.
6. Why is this a good location for marinas?
7. Locate the sports stadium in the photo. Why is this a good location for it?
8. What advantages and disadvantages are there in having an airport in this location?
9. What contrast is there between the foreground and background of this scene?
10. How many things that you see here might be attractive to tourists?
11. Which of the photos, the one above or on p. 88, shows the most peaceful scene? Why do you think so?

ACTIVITY 56
Refer to the photograph above to complete the following:
1. Is this a high or low oblique photograph?
2. Toronto is on the north shore of Lake Ontario. Use this as a guide to establish the four major directions in the photograph.
3. Use the shadow of the C.N. Tower as a help in calculating the approximate time of day the photograph was taken.
4. Describe the general appearance of the land surface in this part of the city.
5. What evidence in the foreground of the photo shows that Toronto is a major transportation centre?
6. (a) The area of tall buildings is the central business core of Toronto. Find this area on the map of Toronto on p. 27 of the atlas.
   (b) Use this map as a guide to name the major expressway in the foreground.
7. Name any five land uses shown in this photograph (e.g. residential, commercial, industrial, recreational, etc.).
8. Name any three patterns in the photo.

Vertical Air Photographs

Vertical photographs, such as the one on p. 93, are taken with the camera pointing directly to the ground as the aircraft flies over it. Because the camera is pointing straight down, tall buildings do not hide objects behind them and there is no foreground or background in the photo. As a result more detail is shown than in oblique photographs. Since you are looking straight down at buildings and land surfaces, objects on the ground may be difficult to recognize. In some ways a vertical photograph is like a map of an area. Maps, however, show only selected information while photographs show everything on the ground. When used together, maps and vertical photographs can help increase your understanding of an area.

ACTIVITY 57
Refer to the photograph on p. 93 to complete the following activity. This photo shows part of the same area located in the oblique photo on p. 91.
1. Establish the directions north, south, east, and west on the photograph.
2. Orient the oblique photo to this one to locate the following: C.N. Tower, central business core, Gardiner Expressway, railway tracks, and grain elevator.
3. Use the shadows to calculate the time of day that the photograph was taken.
4. The use of the grid with the photograph helps to identify the specific locations of various features. Give the grid location for the C.N. Tower.
5. What direction would you travel to get from the C.N. Tower to city hall?
6. Give the grid locations for an example of each of the following:
   parking lot   ship
   railway round house   park

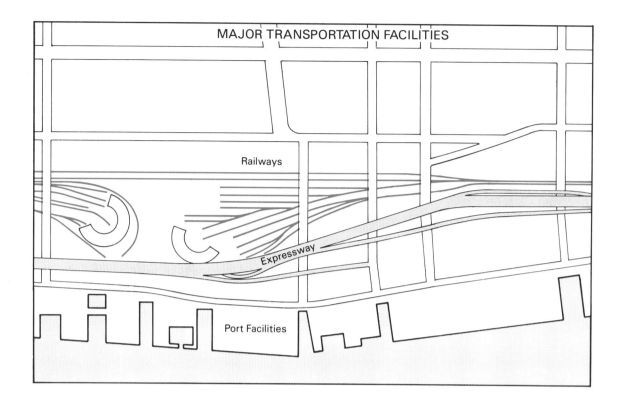

One method of sorting out all the information shown in the photograph is to draw a sketch map of a particular item. Such a map gives you an idea of how that item relates to the whole area. For example, a sketch map of the major transportation facilities would look something like this.

7. (a) Draw a sketch map showing the locations of all the car parking areas.
   (b) Estimate the percentage of the area used for parking.
   (c) Suggest alternative methods of handling the car parking problem.

SIZE AND LOCATION

You now have the skills to refer to the atlas for a great deal of information about topics related to any country or area of the world. As you use the atlas more often, its contents will become familiar and you will find it easier to use.

In order to study a particular topic, you will often need to link together information gained from a variety of maps. The conclusions you draw in this manner are more likely to be remembered than if you had simply read the same information in a text.

Canada's Location in the World

ACTIVITY 58
Refer to the atlas map on pp. 2-3 and more detailed maps on pp. 20-43.
1. On an outline map of Canada label the major water boundaries — Arctic Ocean, Baffin Bay, Atlantic Ocean, St. Lawrence River, L. Ontario, L. Erie, L. Huron, L. Superior, Pacific Ocean.
2. Label our closest neighbours — United States, (Alaska), U.S.S.R., Greenland.
3. Describe Canada's position for each of the following:
   (a) the two hemispheres in which it is located
   (b) the continent on which it is located and its position on that continent
   (c) direction from the main part of the United States
   (d) direction from the U.S.S.R.
   (e) direction from Greenland
   (f) in relation to low, mid, or high latitudes
4. Refer to the following atlas maps: North America, p. 54; South America, p. 55; Europe, p. 82; Asia, pp. 94-95; Australasia, p. 116; Africa, p. 122; Antarctica, p. 160.
   (a) Name the major water boundaries for each continent.
   (b) Name the closest neighbouring continent and the direction you would travel to reach it.
   (c) Name the hemispheres in which each continent is located.

Exact Position

ACTIVITY 59

Refer to the atlas map on pp. 2-3.
1. Name the most northern and southern points in Canada and give the latitude of each.
2. Calculate the range in latitude. (The range is the number of degrees from the most northern to the most southern points.)

   If both locations are on the same side of the equator, find the latitudinal range by subtracting the lowest number of degrees from the highest number of degrees.

$$\text{high} - \text{low} = \text{range}$$
$$86°N - 45°N = 41°$$

   If the locations are on different sides of the equator, find the range by adding them together.

$$\text{lat.} + \text{lat.} = \text{range}$$
$$32°N \to 0° \to 16°S = 48°$$
$$\quad\text{to}\quad\text{to}$$

3. Find the following for each continent listed in Question 4 of Activity 58:
   (a) the most northern and southern points and their latitudes,
   (b) their ranges in latitude.
4. Name the farthest eastern and western points in Canada and give the longitude of each.
5. Calculate the range in longitude. (The range is the number of degrees from the farthest western to the farthest eastern points.)
6. Find the following for each continent listed in Question 4 of Activity 58:
   (a) the farthest eastern and western points and their longitudes,
   (b) their ranges in longitude.
7. List the latitudinal and longitudinal ranges for each continent in a table, and then rank them in order from largest to smallest in north-south and east-west extent.

Size (Distance and Area)

ACTIVITY 60
1. Refer to the map on atlas pp. 2-3 to complete the following:
   (a) Use the map scale to calculate the greatest east-west distance in Canada.
   (b) Calculate the greatest north-south distance in Canada (multiply the latitudinal range by 111 km). Check your answer by measuring the greatest north-south distance and using the map scale.
   (c) Compare the east-west and north-south distances to find which is greater.
2. Refer to the chart on pp. 30A-32A of the atlas to complete the following:
   (a) Find and note the area of Canada.
   (b) Compare the area of Canada to the areas of China, the Union of Soviet Socialist Republics, and the United States.
   (c) Find the total area of all the countries making up North America and calculate the percentage taken up by Canada.
3. Find the following for each of the world's continents:
   (a) the greatest east-west distance,
   (b) the greatest north-south distance.
   (c) Compare the east-west and north-south distances to see which is greater.
   (d) Use the information on atlas pp. 30A-33A to rank order the continents in relation to size.

Refer to the time zone map on atlas p. 138.
4. (a) How many time zones wide is Canada?
   (b) How does Canada compare in time zone width to the U.S.S.R.?
5. (a) How many time zones wide is each continent?
   (b) Rank order the continents in terms of time zone width from most to least.

Boundaries

Some political boundaries are based on parallels of latitude and meridians of longitude. Others are based on natural features.

ACTIVITY 61
1. Refer to the map of Canada on atlas pp. 2-3.
    (a) Name two parallels of latitude and two meridians of longitude that are boundaries and tell what they separate.
    (b) Identify and name five boundaries based on natural features.
    (c) How do boundaries related to parallels and meridians differ from those based on natural features?
2. For each of Europe, Asia, Africa and South America list any five boundaries, tell whether they are based on natural features or meridians and parallels, and tell what they separate.

PROFILES AND CROSS SECTIONS

Two types of diagrams help us to understand the physical appearance of a region. These are known as profiles and cross sections.

PROFILES

A profile is an outline of an object. When that object is a person's head, a profile may look something like this.

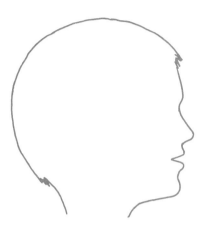

When you look at this profile you get a general impression of a person's appearance. You see the curves of the forehead, nose, lips, and chin. You do not see any detail of eyes, cheeks, or hair. This type of diagram is useful for showing general shape.

When we draw the same kind of diagram for a landscape, we get a general picture of the land surface with no detail to hide it. A physical map such as that on atlas pp. 4-5 uses colour to show the different altitudes of the land surface across Canada. Using the map as a guide, we can sketch a profile that will give a general idea of the land surface across the country. This profile is drawn at approximately the 50°N parallel of latitude.

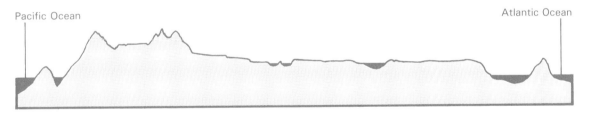

PROFILE ACROSS CANADA AT 50° N LATITUDE

This profile does not give an accurate picture of the land surface. It does give an overall impression of how that surface changes from west to east.

When you draw a profile, include the title, the two end points of the profile, and the location on the map through which the profile is drawn.

ACTIVITY 62
1. Use the map of your province or territory on atlas pp. 20-21, 28-43 as a guide to draw profiles of its physical appearance. (Make these profiles general in nature and avoid the minor changes in altitude.)

(a) north to south
(b) west to east
2. Describe the general nature of the land surface across the area covered by your profile.
3. Refer to the profile of Canada on pp. 4-5 in the atlas.
   (a) Along which parallel of latitude was this profile drawn?
   (b) Identify the major physical features on this profile.
   (c) How many times has the scale of the land elevation been exaggerated over the horizontal scale? What is the effect of this exaggeration?
   (d) Why does the atlas profile show more detail than the one on p. 99 of this book?

CROSS SECTIONS

In some ways, cross sections are similar to profiles. The purpose of the cross section is to show more detail than the profile. While the profile shows only the land surface, the cross section shows what is below the surface. It is widely used to show the geology of land areas.

If we take the profile of Canada and make it into a cross section, it would look something like this.

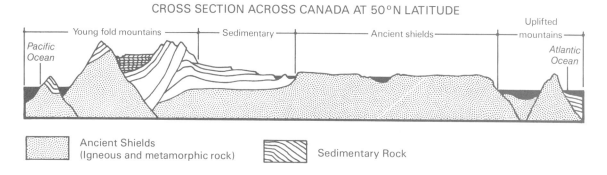
CROSS SECTION ACROSS CANADA AT 50°N LATITUDE

Notice that we not only have the general appearance of the land surface, but also an impression of the underlying rock structure. A legend is necessary to identify the different rock types.

ACTIVITY 63
1. Describe the general nature of the surface of the land from west to east across Canada.

PHYSICAL FEATURES

You can learn a great deal about the physical features of a country or continent by comparing different types of maps.

Landforms

Landforms are the varieties of structures that make up the surface of the earth. The four main types of landforms differ according to the amount of flat and slope land, and to the difference in elevation between high and low points in a given area.

Plains — generally, but not always, areas of low elevation; local relief of less than 200 m is a good criterion for this flat to gently rolling surface; e.g. Gulf Coastal Plain (low altitude), and the Interior Plains (high altitude) of North America

Plateaux — similar to the surface of the plains, although local relief may be greater than 200 m; higher elevations; e.g. Tibetan Plateau (over 3000 m) and southern Africa (about 1000-2000 m high)

Hills — most of the area is sloping and local relief is generally between 200-1000 m; e.g. Swan Hills, Alberta; Cypress Hills, Alberta and Saskatchewan

Mountains — most of the area is sloping and local relief exceeds 1000 m; e.g. Rocky and Andes Mountains

Refer to the map on atlas p. 62 and the cross section diagram on p. 100 of this book to note the underlying rock structure of North America.

ANCIENT SHIELDS

These are the remains of old mountain systems that have been eroded down for millions of years. They are the basement structures on which most continents are built. They consist of a wide variety of rock types. Examples are granite, gneiss, and schist.

ACTIVITY 64
1. (a) Where is this region located in North America?
   (b) Refer to the map of physiographic regions on atlas p. 15 to find the name of this region in Canada.
   (c) How would you describe its shape?
   (d) Refer to the physical map on atlas p. 54 to note general elevations.
   (e) Refer to the photos on p. 106 of this book to describe the surface appearance.
   (f) What minerals are associated with this structure? (In Canada, atlas pp. 15-17; in North America, atlas p. 62.)
2. Refer to the following atlas maps: South America, p. 63; Europe, p. 83; Asia, p. 100; Australasia, p. 116; Africa, p. 127. Answer the following for each of the ancient shield regions shown on these continents.
   (a) Where on the continent is it located?
   (b) What is its shape?
   (c) Refer to the related physical map of each continent to describe the general elevations.
   (d) What minerals are associated with this structure in each continent?

SEDIMENTARY ROCKS

These structures are formed by the cementing together of sediments deposited in ancient seas. Since the ancient shields form the basements of the continents, sedimentary rocks are found lying on top of the shield rocks. (See the cross section diagram on p. 100 of this book.) The types of rock found in sedimentary layers vary according to the types of sediment deposited in the early oceans. Examples are sandstone, limestone, and shale.

ACTIVITY 65
1. (a) Refer to the map on atlas p. 62. Where is this region located in North America?
   (b) What name is given to this region in Canada? (See the map of physiographic regions on atlas p. 15.)
   (c) What is its shape?
   (d) Refer to the physical map on atlas p. 54 to note general elevations.

(e) Refer to the photo on p. 107 of this book to describe the surface appearance.
(f) What minerals are associated with this structure? (See the mining and energy maps on atlas pp. 15-17; the mineral map on atlas p. 62.)

2. Refer to the following atlas maps: South America, p. 63; Europe, p. 83; Asia, p. 100; Australasia, p. 119; Africa, p. 127. Answer the following for each of the sedimentary rock regions shown on these continents.
   (a) Where on each continent is it located?
   (b) What is its shape?
   (c) Refer to the related physical map of each continent to describe the general elevations.
   (d) What minerals are associated with this structure in each continent?

UPLIFTED REMAINS OF ANCIENT MOUNTAIN SYSTEMS

In many respects these are similar in structure and types of rock to the shield areas. The major difference is that forces within the earth have raised these areas to a higher elevation, forming relatively low mountain ranges. (See the cross section diagram on p. 100 of this book.)

ACTIVITY 66

1. (a) Refer to atlas p. 62. Where is this region located in North America?
   (b) What name is given to this region in Canada? (See the map of physiographic regions on atlas p. 15.)
   (c) What is the shape of the region?
   (d) Refer to the physical map on atlas p. 54 to note general elevations.
   (e) Refer to the photo on p. 106 of this book to describe the surface appearance.
   (f) What minerals are associated with this area (atlas pp. 15-17; 62)?

2. Refer to the following atlas maps: South America, p. 63; Europe, p. 83; Asia, p. 100; Australasia, p. 119; Africa, p. 127. Answer the following for each of the uplifted remains of ancient mountain systems shown on these continents.

(a) Where on the continent is it located?
(b) What is its shape?
(c) Refer to the related physical map of each continent to describe the general elevations.
(d) What minerals are associated with this region in each continent?

YOUNGER FOLD MOUNTAINS

Fold mountains are formed when the major plates which make up the surface of the earth slowly move against each other, causing the crust to buckle. (Refer to the map of plate tectonics on atlas p. 140 and the diagram below.) As the crust buckles large sections are lifted high into the air to form rugged mountain ranges. (See the cross section diagram on p. 100 of this book.)

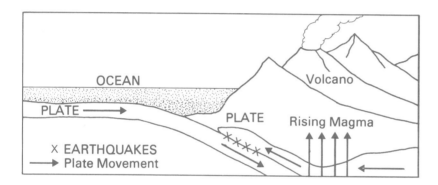

ACTIVITY 67
1. (a) Refer to the map of plate tectonics (atlas p. 140), and earthquakes and volcanoes (p. 141) to see what relationship exists between them.
   (b) What relationship exists between fold mountains and volcanoes?
   (c) Where is the region of fold mountains located in North America?
   (d) What is the shape of the region?
   (e) Refer to atlas p. 15 to see what name is given to this area in Canada.
   (f) Refer to the physical map on atlas p. 54 to note general elevations.
   (g) Refer to the photo on p. 107 of this book to describe the surface appearance.
   (h) What minerals are associated with this structure?

2. Refer to the following atlas maps: South America, p. 63; Europe, p. 83; Asia, p. 100; Africa, p. 127. Answer the following for each of the younger fold mountains regions shown on these continents.
   (a) Where on the continent is this structure located?
   (b) What is its shape?
   (c) Refer to the related physical map of each continent to describe the general elevations.
   (d) What minerals are associated with this structure in each continent?
3. Refer to the maps of world: physical and plate tectonics on atlas pp. 140-141.
   (a) Describe the distribution of mountains on each of the continents.
   (b) Relate the pattern of distribution to the tectonic plates.
   (c) Name the two plates responsible for each of the following mountain systems: Rockies, Andes, Southern Alps, and the Himalayas.

RECENT DEPOSITS

These are loose materials (fine particles of sand, clay, etc.) that are deposited by the wind to form dunes, or by streams to form deltas or alluvium (water-deposited material). Since they are continually being deposited they are much younger than the other structures and are called "Recent."

ACTIVITY 68
1. Where are recent deposits located in North America (refer to atlas p. 62)?
2. Refer to the following atlas maps: South America, p. 63; Europe, p. 83; Asia, p. 100; Australasia, p. 116; Africa, p. 127. Where are the deposits located in each of these continents.

TOP

Ancient shields (vertical aerial photograph)

MIDDLE

Ancient shields (oblique aerial photograph)

BOTTOM

Uplifted remains of ancient mountain systems (oblique aerial photograph)

Sedimentary rocks

Younger fold mountains

River Systems

PATTERNS MADE BY RIVERS

Physical features such as mountain chains and river systems create noticeable patterns on the land surface.

Some of the common patterns of river systems are shown below. These patterns usually result from the nature of the landscape on which the rivers flow.

A. Radial patterns develop on the slopes of mountains.

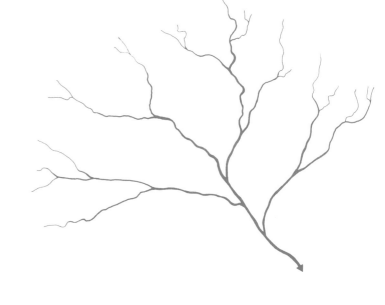

B. Dendritic patterns develop in areas of uniform rock hardness.

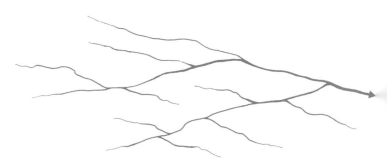

C. Trellis (rectangular) patterns develop in areas of alternate hard and soft layers.

ACTIVITY 69
1. (a) Locate the following rivers in the gazetteer of the atlas: Huang, Yangtze, Ganges, Indus.
   (b) Refer to an atlas map to describe the pattern of each river.

DRAINAGE BASINS

A drainage basin is an area of land drained by a single river system. The Mackenzie River drainage basin includes the Mackenzie River and its tributaries. Find these tributaries on the map on atlas p. 20: Liard River, Peace River, Athabasca River, Redstone River, and Arctic Red River.

All the single river systems that flow into one major water body such as an ocean can be grouped together into larger drainage basins such as the Arctic Drainage Basin, which includes all the main rivers and their tributaries that empty into the Arctic Ocean. Find these on the atlas map on pp. 20-21: Mackenzie River, Horton River, Coppermine River, Hood River, Western River, and the Back River. Drainage basins are separated from each other by watersheds. These are shown by lines drawn on a map that separate the headwaters or sources of different drainage systems. They are sometimes known as divides or heights of land.

ACTIVITY 70
1. Refer to an outline map showing the drainage systems of North America.
   (a) Draw the watersheds to separate the following drainage basins — Arctic Ocean, Pacific Ocean, Atlantic Ocean, Hudson Bay, and the Gulf of Mexico.
   (b) Label the major rivers in each basin.
2. Relate the outline map to the physical map of North America on atlas p. 54. Give a general description of the rivers in each basin (length, speed, description of the course of the river, and uses). Refer to the atlas maps of population on p. 60 and industrialization on pp. 156-157 to describe the present degree of settlement in each drainage basin and to estimate the level of pollution of the major river systems.

SPILLWAYS

At various times during the history of the earth, the climate became cooler. At these times huge ice sheets covered a large part of North America and Europe. (See atlas maps pp. 4 – pleistocene glaciation; 82 – Europe: physical, maximum extent of glaciation; 94-95 – Asia: physical, maximum extent of glaciation.) As these massive, continental glaciers moved across the land, they scraped the northern areas clear of much of their soil, leaving behind bare rock, swampy land, thin, stony soils and thousands of lakes. As the ice melted at the fringes of the glacier, the loose debris carried along was deposited to form a variety of landforms. Among these were the long lines of rolling, hilly land known as moraines.

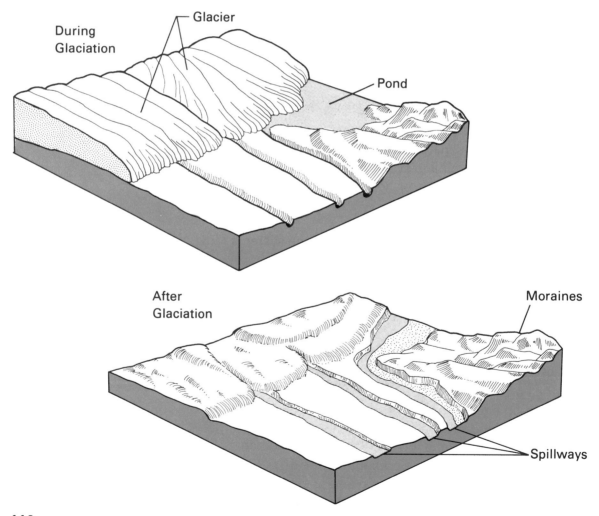

When the climate got warmer again, the southern edges of the continental glaciers melted first. The St. Lawrence River and the Great Lakes were still held locked by the ice. Because these natural drainage channels were blocked, the water accumulated to a height above that of the hills south of the Great Lakes. Gradually the water overflowed these hills, cutting channels as it went. These old channels or stream beds, known as spillways, are broad valleys which may or may not contain rivers today.

The main spillway extended from the site of Chicago to the Illinois River. Other important spillways were:
- (a) Green Bay, Fox River, Lake Winnebago, Wisconsin River to the Mississippi;
- (b) Lake Erie, Maumee River, Wabash River to the Mississippi;
- (c) Lake Erie, Ohio River to the Mississippi;
- (d) Lake Ontario, Mohawk River to the Hudson River.

ACTIVITY 71

Refer to the map on atlas p. 54.
1. (a) Label the major spillways on an outline map of the Great Lakes region.
2. Refer to the following physical maps in the atlas: p. 54 – North America; p. 82 – Europe; pp. 94-95 – Asia.
   (a) Locate the lines showing the maximum extent of glaciation.
   (b) Estimate the percent of each continent that was covered by the continental glacier.
   (c) Locate the areas of the continent where the glacier eroded the topsoil.
   (d) Locate the areas where moraines would have formed.
   (e) Where did glaciers form outside the main continental ice sheet?
3. (a) Label the Mississippi and its major tributaries on an outline map of North America.
   (b) Estimate the percent of the United States occupied by the Mississippi River and its watershed.
   (c) Find the latitude of the most southerly and northerly points of the Mississippi River. Calculate the range in latitude. Remember that parallels of

latitude are about 111 km apart. Multiply the range by 111 km to get an indication of the length of the river. Why is this figure much lower than the actual length?

Refer to the map on atlas pp. 50-51.

(d) Use the colour key as a guide to describe the characteristics of the Mississippi River from its source to its mouth. (Refer to such things as the width and depth of the valley, directness and the speed of the flow, the quantity of sediment, the usefulness of the river for transportation.)

(e) What evidence is there to prove that the Mississippi River carries a high load of sediment?

4. (a) Why is the Mississippi River so prone to flooding?

(b) Evaluate the importance of the Mississippi River and its drainage basin in relation to other drainage basins in North America.

CLIMATE

The weather can be hot and sunny one day and cool and rainy the next. Weather is the day-to-day nature of the atmosphere. When we look at weather over a long period of time, it is known as climate. Look at the maps of January and July temperature and mean annual precipitation on atlas pp. 6-7 to find how the climate in Canada is different across the country. July is used as a sample month because the sun reaches its highest point in the northern hemisphere in June. However, since it takes time for the surface of the earth to heat up, the highest temperatures usually occur in July. In the same way, January is used as a sample month because the sun reaches its lowest point in the northern hemisphere in December. Since it takes time for the earth to cool down, the lowest temperatures usually occur in January.

Temperature Maps

The black lines on the temperature maps on atlas p. 6 are called isotherms. Isotherms are lines joining places having the same average temperatures over a period of time. In the sample below an isotherm has been drawn to join the places having an average temperature of 5°C, 10°C, 15°C, and 20°C.

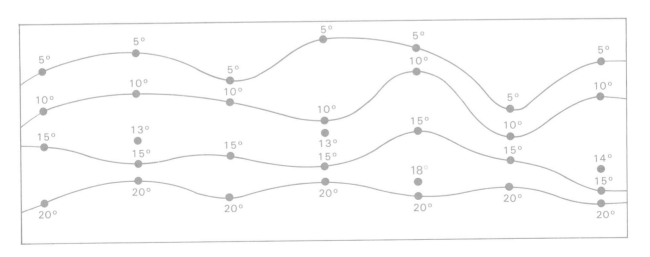

The word isotherm is part of a large family of words all beginning with the letters iso — which means "equal." This family of words names lines on maps joining places with similar quantities. A few of the more common words are:

isotherm	—	temperature
isohyet	—	precipitation
isobar	—	pressure
isobath	—	depth
isoseismic	—	earthquake intensity

To draw July isotherms, for example, you would first mark in the average July temperatures for as many places as possible on the map. Then you would decide on an interval, or how many degrees to leave between each line you will draw. Isotherms are then drawn by connecting all the places that have the same average temperature.

It is possible when using iso-lines to "read between the lines." The intervals between isotherms, for example, create zones on temperature maps. Each zone on the map may be given a different colour, so that the various temperature zones are easily seen.

In the example below the 10° and 20° isotherms have been drawn. This gives an interval of 10°. To draw the 15° isotherm you divide the space between 10° and 20° into equal proportions and mark in the 15° locations.

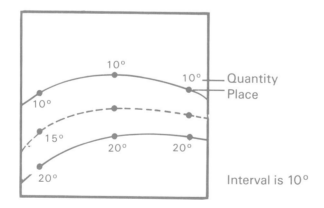

Interval is 10°

ACTIVITY 72

Refer to the January and July temperature maps of North America on p. 56 of the atlas.

1. What are the lowest temperatures in Canada for each of these months? Where are they found?
2. What are the highest temperatures in Canada for each of these months? Where are they found?
3. (a) What is the greatest January temperature range? the greatest July temperature range?
   (b) Which month has the largest temperature range? (Find the range by subtracting the lowest temperatures from the highest temperatures. If a minus reading occurs add the lowest and highest temperatures to find the range.)

   high   -   low   =   range
   60°C - 20°C = 40°C

   high   +   low   =   range
   45°C → 0° → -20°C = 65°C
         to   to

4. Refer to the following temperature maps in the atlas:
   North America, p. 56
   South America, p. 57
   Europe, pp. 80-81
   Asia, pp. 96-97
   Africa, p. 123
   Australasia, p. 117

   Draw and complete a chart similar to that below for January and July temperatures.

CONTINENT	HIGHEST		LOWEST		RANGE	
	Jan.	July	Jan.	July	Jan.	July
North America						
South America						
Europe						
Asia						
Africa						
Australasia						

5. Which one of the listed continents has the greatest January temperature range? the greatest July temperature range?

High and Low Pressure Systems

Wind is air in horizontal motion, caused by air pressure differences that are due to unequal heating of the earth's surface. When air becomes heated (as over a large land mass in summer) its air pressure decreases. When air cools (as over a large land mass in winter) its air pressure increases. Air flows from high to low pressure areas to even out the differences. Because of the earth's rotation, winds do not blow directly north or south towards the low pressure zones. They are deflected to the right in the Northern Hemisphere and to the left in the Southern Hemisphere.

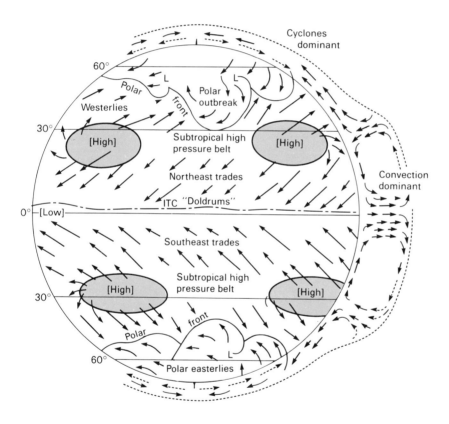

This diagram of surface winds does not show the disrupting effect of the large continents in the Northern Hemisphere. "ITC" stands for the Intertropical Convergence Zone, where the trade winds meet and cause the air to rise. "Doldrums" refers to a belt of calms and variable winds formed at certain times of the year when the trades do not come together.

ACTIVITY 73

Refer to the maps of pressure and winds on atlas pp. 142-143.
1. (a) What differences are there between the January and July maps and the diagram on this page?
   (b) Which of these months has the greatest difference?
2. What is the cause of these differences?
3. What happens to the pressure and wind systems as the seasons change?
4. Which two wind systems are the most constant? Are they more constant in January or July?

Ocean Currents

Surface currents in the oceans are set in motion by prevailing winds. The earth's rotation causes these currents, like winds, to be deflected to the right in the Northern Hemisphere and to the left in the Southern Hemisphere. However, with the friction between air and water, surface currents move in a direction about 45° farther right (Northern Hemisphere) or left (Southern Hemisphere) of wind direction.

Other minor factors influencing current direction include differences in water density due to temperature and salinity, and the shape of ocean basins and coasts.

ACTIVITY 74

Refer to the maps of pressure and winds and ocean currents on atlas pp. 142-143.
1. What currents are associated with the following winds: North East Trades, Westerlies, South East Trades?
2. Does the seasonal shift of the wind patterns have much effect on the pattern of ocean currents?

117

CLIMATE CONTROLS THAT AFFECT TEMPERATURE

When you work with climate maps there are two main steps to follow. The first is to look at the maps and search for general patterns. The second is to explain the reasons for these patterns. Climate controls are conditions on the earth that help explain climate patterns.

Latitude

Generally, the farther north or south of the equator, the cooler the temperature will be. The main reason for this is seen in the diagram.

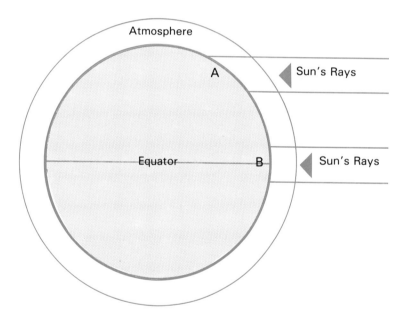

Equal amounts of the sun's energy are affected as follows:

Point A — Sun's rays cover a large area and pass through more atmosphere, which results in lower temperatures.

Point B — Sun's rays cover a small area and pass through less atmosphere, which results in higher temperatures.

The amount of heat received at any place is also affected by the movement of the earth as it revolves around the sun. This movement is responsible for seasonal changes.

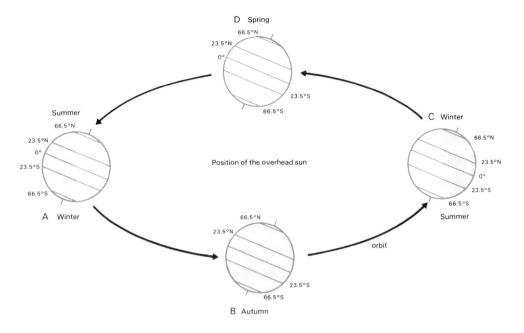

The earth is tilted at an angle to the sun. Because of this the amount of heat received at any place will vary during the year. The northern hemisphere is tilted towards the sun in position A. Places in that hemisphere are receiving more direct rays of the sun for longer periods than places in the southern hemisphere. Therefore the northern hemisphere receives more heat. In position C, the southern hemisphere is tilted towards the sun. As a result it receives more heat than the northern hemisphere. In positions B and D the sun's rays reach both poles, and both hemispheres receive the same amount of energy from the sun.

In position A, while the northern hemisphere is enjoying summer, places south of the equator are having their winter season. In position C, while southern hemisphere places are enjoying their summer, we are having our winter. This is known as season reversal.

Altitude

The higher the elevation above sea level, the cooler the temperature will be. This happens because the earth's atmosphere acts like the glass on a greenhouse. The sun's rays pass through the atmosphere, warming it very little. When the rays reach the earth they are absorbed and the earth heats up. This heat is then sent back into the atmosphere, warming it up. Thus the higher one goes in elevation, the farther one gets from the source of heat — the earth. Remember that low temperatures in unexpected places may be caused by altitude. For example Mexico City, which is at an altitude of 2282 m, and only 19°N of the equator, has a July temperature of 17°C.

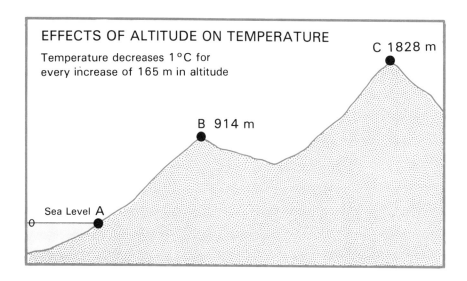

EFFECTS OF ALTITUDE ON TEMPERATURE
Temperature decreases 1°C for every increase of 165 m in altitude

ACTIVITY 75
1. If the temperature at A is 27°C, calculate the temperature at B; at C.
2. If the temperature at C is 6°C, calculate the temperature at B; at A.
3. Use the gazetteer to locate the following places. Note the latitude and longitude of each place. Calculate the temperature at each place, if the sea level temperature is 30°C.
   (a) Mt. Eduni
   (b) Aling Kangri
   (c) Communism Mt.
   (d) Mt. Everest
   (e) Damavand
   (f) El'brus
   (g) Mont Blanc
   (h) Tatry

Water
Large areas of water affect the temperature of the nearby land. They make a place cooler in summer and milder in winter (this is called moderating the temperature) than places farther inland. Places where

this effect is most noticeable have marine climates. Inland areas far away from water will usually have colder winters and hotter summers. Places where this effect is most noticeable have continental climates.

Ocean Currents

If the temperature of an ocean current is warmer than the body of water in which it flows, it is called a warm current. A cold current has a lower temperature than the body of water through which it flows. Ocean currents affect the temperature of the air blowing over them. If that air then blows to the land, the temperature of the land will be changed. For major ocean currents, see the maps on atlas pp. 142-143.

EFFECT OF LARGE AREAS OF WATER ON TEMPERATURE

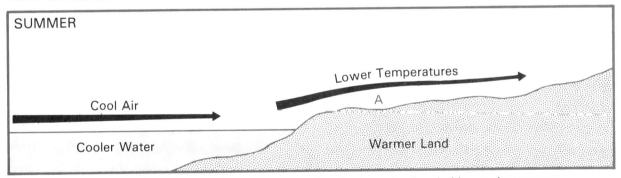

SUMMER: Large bodies of water heat up more slowly than the land. Air moving across the water becomes cool and lowers the temperature of places located near the coast.

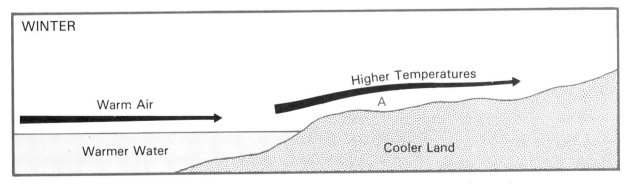

WINTER: Large bodies of water cool down more slowly than the land. Air moving across the water becomes relatively warm and raises the temperature of places near the coast.

Prevailing Winds

Winds blowing from large areas of water to the land will carry the moderating effect of the water inland. Winds blowing from the land to the water will reduce the moderating effect of coastal waters. To find the major winds affecting an area look at the maps on atlas pp. 142-143.

ACTIVITY 76

Refer to the January and July temperature maps of North America on p. 56 of the atlas.
1. (a) In what general direction do the isotherms run in each of January and July?
   (b) Where are the most noticeable changes from these general directions found?
   (c) If the surface of the earth consisted only of land at a uniform height, in what direction would all isotherms run? What climatic control would cause this pattern?
   (d) Use the climate controls described previously to explain the January and July isotherm pattern in Canada.
2. Select the one climatic control that is dominant in Canada for both January and July.

ACTIVITY 77

Refer to the following temperature maps in the atlas:
North America, p. 56
South America, p. 57
Europe, pp. 80-81
Asia, pp. 96-97
Africa, p. 123
Australasia, p. 117
1. Note the following for each of January and July in North America, South America, Europe, Asia, Africa, and Australasia:
   (a) the general direction of isotherms
   (b) the locations of areas where the greatest variation from this general direction occur
   (c) the climate controls causing these variations
2. (a) For each continent note whether the January or July isotherm pattern shows the greatest variation from normal.

(b) Select the one climatic control for each continent that is dominant in January and the one that is dominant in July.

Mean Annual Precipitation Maps

Precipitation is made up of either liquid or solid forms of water which reach the earth from the air. These forms are rain, sleet, snow and hail.

The black lines on precipitation maps are called isohyets. Isohyets are lines joining places having the same amount of precipitation over a period of time. In the sample below an isohyet has been drawn to join places having a mean precipitation of 200 mm and 300 mm.

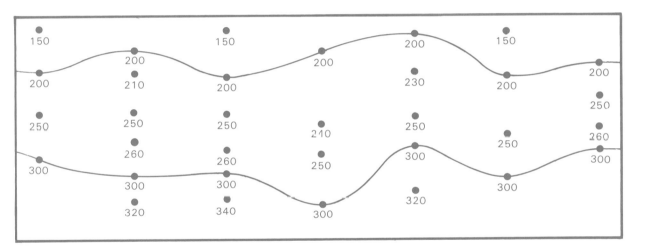

ACTIVITY 78
Refer to the map of mean annual precipitation in Canada on p. 7 of the atlas.
1. What is the heaviest precipitation in Canada? Where is it found?
2. What is the lowest precipitation? Where is it found?
3. Calculate the range in precipitation.
4. Refer to the following mean annual precipitation maps in the atlas: North America, p. 56; South America, p. 57; Europe, p. 80; Asia (May to October), p. 96; Australasia, p. 116; Africa, p. 123. Draw and complete a chart for mean annual precipitation similar to the one shown.

CONTINENT	HIGHEST	LOWEST	RANGE
North America			
South America			
Europe			
Asia (May to Oct.)			
Australasia			
Africa			

CLIMATE CONTROLS THAT AFFECT PRECIPITATION

There are three main causes of precipitation. One of these is known as relief or orographic precipitation. (Orography means dealing with mountains.)

This type of precipitation results in heavy rains on the windward side of mountains (see diagram below). Very little rain falls on the leeward side of the mountains. The leeward side is said to be in the "rain shadow" of the mountains.

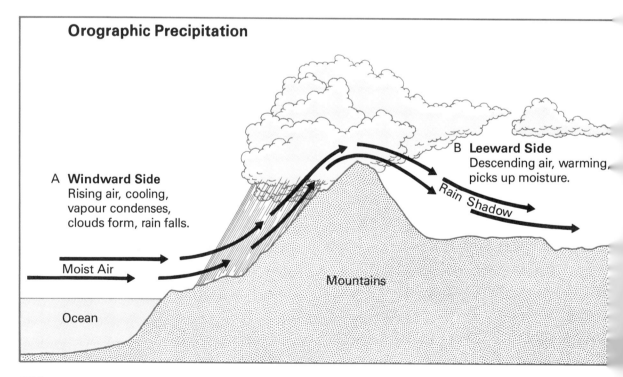

Orographic Precipitation

A **Windward Side** Rising air, cooling, vapour condenses, clouds form, rain falls.

B **Leeward Side** Descending air, warming, picks up moisture.

The second type of precipitation is called convectional, because it is caused by convection currents in the atmosphere.

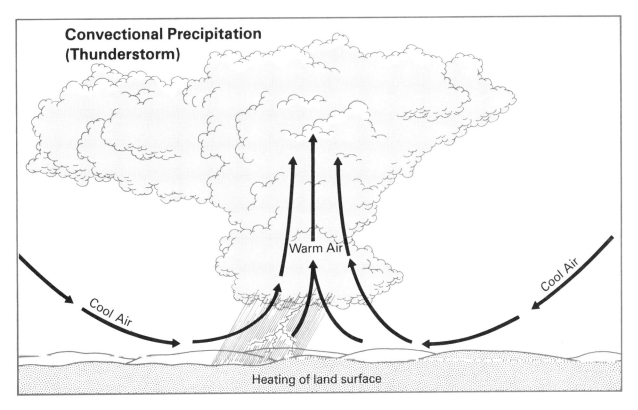

Convectional Precipitation (Thunderstorm)

These currents are created when the weather gets hot. As the land heats up, it warms the air above it. The warm air rises and is replaced by surrounding cooler air which heats up and also rises. As the warm air rises, it cools, and the water vapour in it condenses to form clouds and rain or hail.

These types of storms cover a relatively small area, last a short time, and produce a heavy precipitation. Because they are a result of hot weather, they are a major cause of precipitation in the tropics and some summer precipitation in more northerly latitudes.

Convectional precipitation may also be caused by the movement of air masses.

The third type of precipitation is known as cyclonic or frontal. Over the continent of North America large masses of air are continually in motion. An air mass is simply a huge body of air in which the temperature and humidity are fairly uniform all over. When two air masses with different characters (hot and humid from the Gulf of Mexico vs. cold and dry from the Arctic) meet, a battle begins in the atmosphere. Air mass movement is frequent over much of North America, and most of Canada's precipitation is cyclonic.

Cyclonic Precipitation (Low Pressure Storms)

Warm air rises quickly in front of the steep-edged, fast-moving cold front (a front is a line separating two different air masses). This is another cause of thunderstorms. Strong winds, a drop in temperature, and heavy precipitation may be experienced as the cold front moves through an area. Temperatures will be lower after the front has gone by.

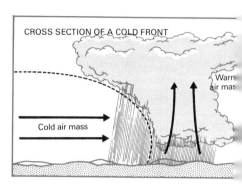

Warm air rises slowly over the cooler air in front of it. Cloud cover and rainfall will be heavier and continue for a relatively long time (many hours or days). Temperatures are higher after the warm front has passed through.

ACTIVITY 79

Refer to the maps of air masses and winds on p. 6 of the atlas.
1. (a) What names are used to show whether the air masses originate over water or land?
   (b) What is the main difference between air masses originating over water and those originating over land?
2. Why is the air mass over the Arctic islands of Canada labelled as continental in January, but maritime in July?
3. Which air masses dominate Canadian weather most of the year?
4. Which air mass brings most of our warm weather in July?
5. Which two air masses would be responsible for most of the cyclonic circulation over much of Canada?

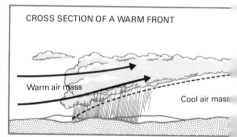

ACTIVITY 80
Refer to the mean annual precipitation map of North America on p. 56 of the atlas.
1. (a) In what general direction do the isohyets run?
   (b) How does this compare with the general direction of the isotherms on the temperature maps on p. 6?
   (c) Where are the most noticeable changes from this general direction?
   (d) Account for the pattern of isohyets on this map.
2. Select the one climate control that is dominant in Canada for mean annual precipitation.
3. Refer to the following mean annual precipitation maps in the atlas: North America, p. 56; South America, p. 57; Europe, p. 80; Asia (May to October), p. 96; Australasia, p. 116; and Africa, p. 123. For each continent note:
   (a) in what general direction the isohyets run;
   (b) how this compares with the general direction of the isotherms;
   (c) where the most noticeable changes from this general direction are.
   (d) Account for the pattern of isohyets on these maps.
4. Select the two most significant factors in the distribution of precipitation in each continent. Explain the reasons for your choice.

Climate Graphs

The climate graphs on p. 7 of the atlas give detailed information about specific places in Canada.

Each climate graph is made up of two major parts. The first is a line graph for temperature.

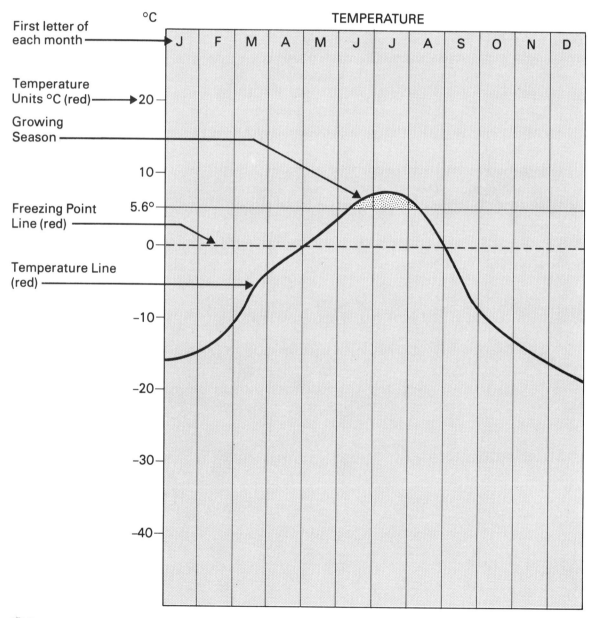

LINE GRAPHS

The range of temperature is easier to see on a line graph than from a set of statistics. When you construct a line graph, the horizontal axis is used to show the time period (months), and the vertical axis is used to show the temperatures.

The first step in drawing a line graph for temperature is to find the average temperature for each month. These are then plotted on the graph and connected with a line. Edmonton will be used as an example. The temperatures for Edmonton are:

	J	F	M	A	M	J	J	A	S	O	N	D
°C	-15	-11	-5	4	11	15	18	16	11	5	-4	-11

When these are plotted on the outline of a graph we have:

When joined together they form a completed line graph.

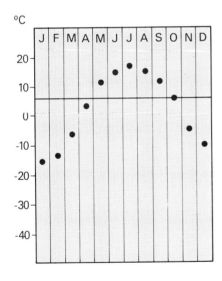
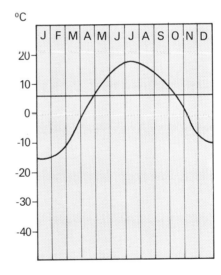

129

Note the shape of the temperature line in each of the three climate graphs below.

Northern Hemisphere

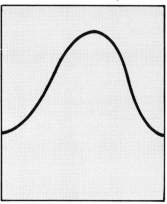

The line points to the top of the graph.

Tropics

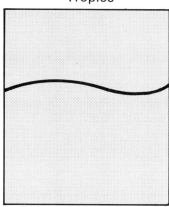

A roughly straight line.

Southern Hemisphere

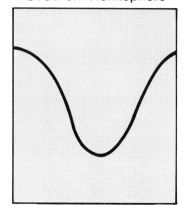

The line points to the bottom of the graph.

To find out why the lines have these shapes, refer back to p. 119 of this book.

Examine the climate graphs on atlas pp. 55 and 57 for the following six places:
Revelstoke and Yuma
Manaus and Bogota
Punta Arenas and Buenos Aires

Find each of these places on the map. Compare the temperature line on each climate graph with the diagrams above. Does the shape of the temperature line match the pattern above for each hemisphere location?

ACTIVITY 81

1. Draw line graphs to show the temperatures of these cities.

Ottawa

	J	F	M	A	M	J	J	A	S	O	N	D
°C	-11	-11	-4	5	13	18	21	19	14	7	0	-9

Belem, Brazil

	J	F	M	A	M	J	J	A	S	O	N	D
°C	26	26	27	27	27	27	26	26	27	27	27	27

Valdivia, Chile

	J	F	M	A	M	J	J	A	S	O	N	D
°C	17	17	15	12	10	8	8	8	10	12	13	15

2. Refer to the climate graphs on pp. 54-57 of the atlas.
   (a) How can you tell the month when the letters are not included on the graph?
   (b) How can you tell the months on the graphs of Wadi Halfa (atlas p. 123), and Antofagasta (atlas p. 57)?
3. (a) Find the latitude and longitude of each of the places in Question 1.
   (b) How does the temperature graph give some idea about general location in the hemisphere?
   (c) Is this general location related to the latitude or longitude?
4. What major climate controls are responsible for the temperatures of these three places?
5. Refer to the climate graphs on atlas p. 7. The green shaded areas of these graphs show the growing season.
   (a) When does the growing season start?
   (b) Which graph shows the longest growing season?
   (c) Which graph shows the shortest growing season?
   (d) Calculate the approximate number of days of growing season for Yellowknife, Vancouver, Winnipeg, and Halifax.
   (e) Name the two months which are fully included in every one of these graphs.
6. Refer to the growing season map on atlas p. 6.
   (a) What do the lines on this map show?
   (b) In what areas of Canada are the longest (more than 220 days) and shortest growing seasons located?
   (c) What climatic control accounts for the length of the growing season on the west coast?
   (d) What climatic control accounts for the overall pattern shown by this map?
   (e) How long is the growing season in the area in which you live?
   (f) Locate the position of Yellowknife, Vancouver, Winnipeg, and Halifax. How close were your answers in Question 5 (d) to the information shown on the map?

The second major part of climate graphs is the bars showing precipitation.

BAR GRAPHS

In many respects bar graphs are similar to line graphs. They are relatively easy to construct and easy to interpret. On the graph below, the horizontal axis is used to show the time period (months), and the vertical axis is used to show the precipitation quantities. Because all the bars represent precipitation, they are joined together to show a continuity from one month to the next.

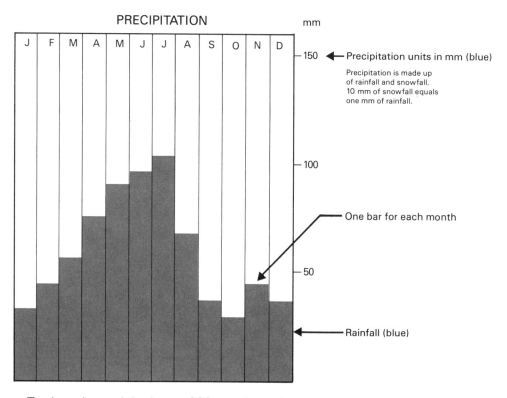

Total yearly precipitation — 630 mm Annual

To draw the bars showing precipitation you must have the average precipitation figure for each month.

Edmonton

	J	F	M	A	M	J	J	A	S	O	N	D
mm	25	20	17	23	37	75	83	72	36	19	19	21

These figures include snowfall and rainfall together, so everything will be shown as rainfall.

These are marked on the graph by drawing a straight line across the column for each month.

When they are shaded in we have a picture of the precipitation pattern.

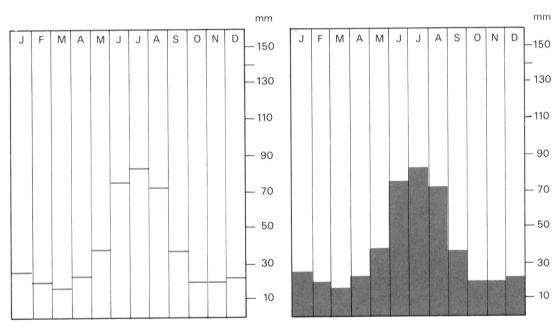

477 mm Annual

Sometimes when the precipitation is very heavy the scale will be extended above the graph so that all of it can be shown. The graph of Cherrapunji on atlas p. 97 is a good example of this.

ACTIVITY 82

Draw precipitation graphs for these cities using the same scale for each. Make sure the scale accommodates the largest amount of precipitation in these figures.

Ottawa

	J	F	M	A	M	J	J	A	S	O	N	D
mm	74	56	71	69	64	89	86	66	81	74	76	66

Belem, Brazil

	J	F	M	A	M	J	J	A	S	O	N	D
mm	318	358	358	320	259	170	150	112	89	84	66	155

Valdivia, Chile

	J	F	M	A	M	J	J	A	S	O	N	D
mm	66	74	132	234	361	450	394	328	208	127	124	104

After you have drawn the graphs, add the monthly precipitation figures and place the total at the bottom of each graph.

ACTIVITY 83

1. Why is it better to extend the graph of precipitation for Cherrapunji than to increase the numbers in the scale to accommodate the large amounts of precipitation?
2. Why must you be careful when comparing quantities of precipitation on the graphs for Canada on atlas p. 6 with graphs for other places shown elsewhere in the atlas?
3. Is the diagram of orographic precipitation on p. 124 of this book a profile or cross section? Why is the one used the best for this diagram?

When the temperature line graph and the precipitation bar graph are put together they form a climate graph.

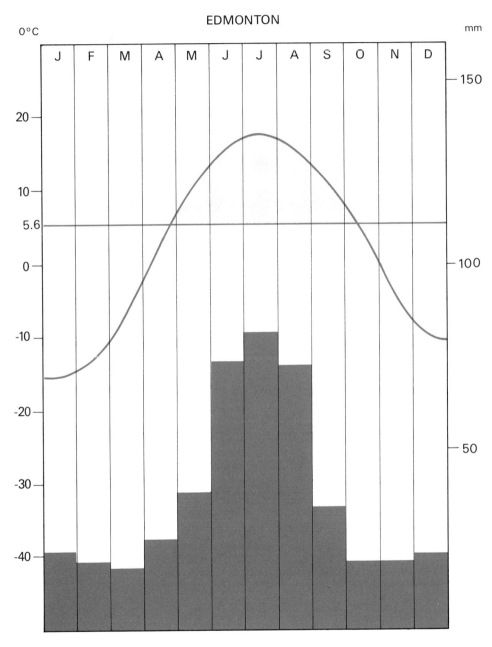

447 mm Annual

INTERPRETATION OF CLIMATE GRAPHS

These graphs are useful only if you know what to look for. The following questions indicate the kind of information you can find from a climate graph. Answers are based on the graph for Edmonton.

General

Questions   *Answers*

(a) At what latitude is the place located?   northern hemisphere, 54°N

(b) What is its elevation?   676 m

Temperature

Questions   *Answers*

(a) What is the highest temperature?   highest 18°C (warm)
   In which month does it occur?   month — July

(b) What is the lowest temperature?   lowest −15°C (very cold)
   In which month is it found?   month — January

(c) What is the range in temperature?   18° to 0° to −15° = 33°C (very large)

Precipitation

Questions   *Answers*

(a) Calculate the yearly total.   yearly total — 447 mm (light)

(b) Is the precipitation fairly evenly spread over the year, or is there a seasonal maximum and minimum? (Remember to be careful of the reversal of seasons in the Southern Hemisphere.)   There is a summer maximum (mainly in June, July, and August).

(c) In which months would the precipitation probably be mostly in the form of snow?   Use the months in which the temperature averages below 0°C. This is not accurate, but gives a general idea. — 5 months, Nov., Dec., Jan., Feb., March

(d) 10 mm of snowfall = 1 mm of rainfall. Use this figure to calculate the average yearly snowfall in Edmonton.

This will give some idea of amounts, but is not accurate.
Snowfall
Sept. — mm 0.5 x 10 = 5.0
Oct.   —   10 x 10 = 100
Nov.   —   20 x 10 = 200
Dec.   —   25 x 10 = 250
Jan.   —   25 x 10 = 250
Feb.   —   20 x 10 = 200
Mar.   —   20 x 10 = 200
Apr.   —   15 x 10 = 150
May   — mm 0.5 x 10 = 5.0
Total: 1360.0 mm (heavy)

(e) Does the highest precipitation occur at the same time as the highest temperature? (This is important for agriculture.)

Yes. Crops will receive the most precipitation at the same time they receive the greatest amount of heat.

All the above questions do not need to be asked about each climate graph. Choose from these questions the ones that help develop the topic under study.

When giving word descriptions of numbers, the following will be of help:

(a) *Temperature*
under −10°C   very cold
−10°C to −1°C   cold
0°C to 9°C   cool (mild in winter)
10°C to 19°C   warm
20°C to 30°C   hot
over 30°C   very hot

(b) *Temperature Range*
under 5°C   small
6°C to 15°C   moderate
16°C to 30°C   large
over 30°C   very large

(c) *Annual Precipitation*
under 250 mm   sparse
250 mm to 499 mm   light
500 mm to 999 mm   moderate
1000 mm to 1999 mm   heavy
over 2000 mm   very heavy

ACTIVITY 84
Estimate the position of Edmonton on the mean annual precipitation map and on the two temperature maps on pp. 6-7 of the atlas.
1. Compare the colour symbols on the maps with the information on the climate graph of Edmonton on p. 135 of this book. Do the maps give the same information as the climate graph?
2. What advantages are there to using climate graphs instead of temperature and precipitation maps?

CLIMATE CONTROLS FOR EDMONTON

Now that we know what the climate of Edmonton is like, the second major task is to see what makes the climate the way it is. We want to see which climate controls affect the climate of this city.

Temperature
1. *Latitude* — Edmonton has a latitude of 54°N. This places it towards the northern part of the mid-latitudes. Low temperatures should be expected.
2. *Altitude* — The maps of urban plans on atlas p. 27 show a number of Canadian cities. Under each set of climate statistics the altitude is also shown. The altitude of Edmonton is 676 m. Since temperatures decrease 1°C for every increase of 165 m in altitude, this means that Edmonton's temperatures are approximately 4°C lower than they would be at sea level. This is not a very important control.
3. *Water* — Edmonton is well inland from large bodies of water, and the Rocky Mountains act as a barrier, so Edmonton has a continental climate with extremes of temperature. This is shown by the very large temperature range of 33°C.
4. *Prevailing Winds* — A look at the maps of winds on atlas pp. 142-143 shows that the Westerlies affect Edmonton in January and July. These winds do not moderate the temperatures because they have lost their moderating effect on the Pacific side of the Rocky Mountains.

Precipitation
5. Much of the precipitation is cyclonic, caused by the meeting of cold and warm air masses. Since the greatest amount of rain falls during the warmest months a certain amount will also be convectional.

In summary, the most important controls affecting the climate of Edmonton are:
    (a) its latitude
    (b) its continental position
    (c) cyclonic storms

ACTIVITY 85

To complete your study of climate, compare the graphs for: Prince Rupert, B.C., atlas p. 55; Alert, atlas p. 54; and Halifax, Nova Scotia, atlas p. 7. (Remember that the climate graphs on p. 7 have a different scale for precipitation than the scales used in the rest of the atlas.)

1. Copy and complete a chart like the one below.

Climate	Prince Rupert	Alert	Halifax
Temperature			
(a) highest and month			
(b) lowest and month			
(c) range			
Precipitation			
(a) yearly total			
(d) distribution			
(c) main snow months and approximate total			
(d) relationship of high temperature to high precipitation			

2. Copy the chart below and list the *one* major control having the greatest effect on temperature and precipitation at each place.

Controls	Prince Rupert	Alert	Halifax
(a) temperature			
(b) precipitation			

3. Develop and complete charts similar to those in Questions 1 and 2 for Ottawa, Belem, and Valdivia, using the climate graphs you have already drawn for these locations.

Climate Regions

A region is an area of the earth which can be identified by specific characteristics. If you take all the information shown on the maps on atlas pp. 6-7 and average the temperatures and precipitation over long periods of time, it becomes possible to divide Canada into climate regions as illustrated on the map on atlas p. 8. The temperature and precipitation in each region is fairly uniform over the whole area. Remember that the boundary lines should be regarded as zones of transition rather than lines of complete change. This means that the changes from one region to the next are gradual rather than sudden.

ACTIVITY 86
Refer to the map of climate regions on p. 8 of the atlas.
1. Find the climate region for each of the following settlements: Prince Rupert, Alert, and Halifax.
2. (a) Compare the legend descriptions for each location with the description you made for each location in Activity 85.
   (b) How closely do they resemble each other?
   (c) What are the main differences?
   (d) Why would you expect to find some differences between the two sets of information?
3. (a) Compare the estimated snowfall for Edmonton on p. 137 of this text with the amount shown on the snowfall map on p. 7 of the atlas.
   (b) Account for any differences.
4. Refer to the climate graphs of Bogota (atlas p. 57) and Singapore (atlas p. 97).
   (a) What are the latitudes of each?
   (b) Account for the differences in temperature.
   (c) Account for the differences in precipitation.

NATURAL VEGETATION

Natural vegetation is the plants that grow undisturbed by people.

Vegetation maps like the ones on atlas pp. 9 (Canada) and 84 (Europe) show the zones covered by each main type of vegetation. On maps like these, a clear boundary line separates different kinds of vegetation. In actual fact there is a gradual change from one main type of vegetation to another. However, map makers must decide where to draw the boundary lines. To do this they choose a place in between two clearly different zones where plants of both types are mixed together. The diagram below shows how this is done.

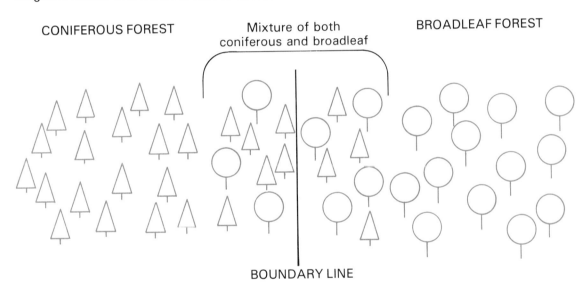

ACTIVITY 87

Refer to the vegetation map on p. 9 of the atlas.
1. List the vegetation zones in Canada.
2. Research and draw a chart to describe two zones and the main uses made of the plants.
3. Refer to the map of January temperature on p. 6 of the atlas. Calculate which isotherm is most closely related to the southern boundary of the coniferous forest region.
4. Locate the boundary line between the prairie and open woodland regions. Relate this to the map of mean annual precipitation on atlas p. 7 to indicate

approximately how many millimetres of precipitation are necessary for tree growth.
5. The climate of an area has a direct effect on vegetation. Locate the climate graph of Aden on p. 96 of the atlas. How is the total amount of precipitation reflected in the vegetation region in which it is located?
6. Find the relationship between temperature, precipitation, and vegetation for the following places. The numbers in parentheses refer to atlas pages.
   (a) Arkhangel'sk (pp. 96, 98)
   (b) Singapore (pp. 97, 98)
   (c) Cherrapunji (pp. 97, 98)
   (d) Barcelona (pp. 80, 84)
   (e) London (pp. 80, 84)
   (f) Frobisher Bay (pp. 54, 63)
   (g) Rio de Janeiro (pp. 57, 63)
7. Refer to the vegetation map of Canada on p. 9 of the atlas. Use the Canada-U.S. boundary as a base line to draw a profile of vegetation zones from the Pacific Ocean to Lake Superior.

ACTIVITY 88
Refer to the maps of climate regions and vegetation on atlas pp. 8 and 9.
1. Find three climate regions which have boundaries fairly similar to three vegetation regions and use the map keys to describe their relationship.
2. Refer to the growing season map on atlas p. 6. What number of days correlates closely with the boundary between open woodland and tundra?
3. Make a general statement about the relationship between climate and vegetation.

A coniferous forest

A broadleaf forest

A mixture of broadleaf and coniferous trees

143

A desert

A tropical rainforest

Temperate grassland

Tropical grassland

ACTIVITY 89
Refer to the vegetation maps on the following pages of the atlas: North America and South America, p. 63; Europe, p. 84; Asia, p. 98; Australasia, p. 118; Africa, p. 124.
1. (a) Which two of these maps show the greatest variety of vegetation?
   (b) Which two show the least?
2. Name the region which is found only on the map of Europe.
3. Which area is the only one having no hot desert?
4. Which continent has the most symmetrical distribution of vegetation?
5. What is the relationship between tropical rainforest and annual rainfall?

AGRICULTURE

Farmers raise animals and grow crops on their farms to produce our food and some of our clothing. We call the work of farmers agriculture. Agricultural activity is very important in Canada. It supplies not only the food we eat but also the food we export to other countries. Wool from sheep and leather from cattle are used for some of our clothing.

Many people not living on farms have jobs related to agricultural activities. Meat-packing plants, wineries, flour mills, woollen mills, shoe factories, and dairies are only a few that rely on farm products. Farmers, on the other hand, purchase many kinds of machinery and supplies to operate their farms. In this way, jobs are created in the manufacturing centres.

ACTIVITY 90
1. (a) Find the percentage of the total land area in Canada suitable for agriculture. (Look under the heading "Arable and Permanent Pasture" on atlas p. 30A. Arable means "land that can produce crops." Even if the land is presently not farmed, it is still classed as arable.)
   (b) Calculate the amount of arable and pasture land in Canada in 1980, using the answer you get from (a).
2. (a) Calculate the amount of arable and pasture land per capita (for each Canadian).
   (b) Make the same calculation for the United States, which has an area approximately that of Canada. Compare the Canadian and American figures.

$$\frac{\text{Arable Land + Pasture Land}}{\text{Population}}$$

Factors that Affect Farming

In spite of Canada's large land area, a number of factors limit the amount of its farmland.

(a) **Climate** — The right climate is essential for farming. This means that the combination of temperatures and precipitation must be suitable for the crops being grown. In general, warmer temperatures are more favourable for crop production. The length of the frost-free season is very important in farming; that is the number of days between the last frost in spring and the first frost in fall. The longer this period of time, the greater chance crops have to mature without the danger of being killed by frost.

The total amount of precipitation and its distribution during the year are also important factors. The best combination is to have the highest amount of precipitation during the warmest months. This will help ensure the growth and development of young plants into mature crops.

There is no single answer to what is a suitable climate for crops. Each crop has its own particular needs. For example, what is good for bananas is not good for wheat.

Climate hazards are a continuing threat to farmers. Such things as summer drought, hailstorms, severe rain and windstorms, late spring frosts and early fall frosts can wipe out a crop in a very short time. All of these make farming a risky business.

(b) **Landforms** — Many of the landforms in Canada are not suitable for farming. The young fold mountains in western Canada consist of bare rock surfaces, steep slopes, high altitudes, and little soil except along river valleys or plateaux.

The ancient mountain systems of eastern Canada are lower in altitude but the land surface is still rugged. There is much bare rock. Most of the soil is generally found along river valleys and coastal plains.

The ancient shield, which makes up the largest single part of Canada's land surface, has been scraped bare of soil in most places by the movement of continental glaciers. A large part of the shield consists of bare rock, lakes, and swamps. Farming is possible only in those scattered patches that were once ancient lake beds.

The best areas for farming in Canada are the extensive sedimentary rock areas of the western interior, southern Ontario, and the St. Lawrence river valley. These areas make good farmland because they have deposits of soil and a relatively flat land surface.

(c) **Soils** — There are different types of soils, and some types suit some crops better than others. For example, the sandy soils suitable for growing tobacco would not be suitable for growing wheat.

The most important factor about the soil is its fertility. Soil fertility is a combination of two things — its mineral content and its humus (the amount of organic matter formed from decayed vegetation). Of lesser importance are its texture (the size of soil particles, which affects its ability to hold moisture), and its depth (from the surface to bedrock).

(d) **Markets** — Farmers must be able to sell their crops and animals to earn a living. If people want to eat more of one food and less of another, farmers must be prepared to change the products they raise on their farms in order to be able to sell them. For example, if more people start eating more eggs and poultry, farmers might be encouraged to keep more chickens.

(e) **Natural hazards** — Other than the climatic hazards referred to previously, farmers are also faced with the problems of natural pests such as army worms or grasshoppers, and the many diseases that affect both plants and animals.

ACTIVITY 91
1. Which two of the above factors affecting agriculture would set the greatest limits on farming in Canada?
2. Refer to the agriculture map on p. 12 of the atlas to name the major agricultural regions of Canada. List the two main factors influencing the location of each region.
3. According to the information on atlas p. 11A, which province had the greatest amount of farmland in 1981? Which province had the least?

Wheat Farming

Wheat takes up the largest area of land of any field crop grown in Canada. Canada ranks as one of the world's greatest wheat producers. Wheat is grown in most provinces in combination with other crops and livestock. But there is one area where wheat is the major product by itself.

ACTIVITY 92
1. Find Canada's positions as a world wheat producer and exporter using the circle graphs on p. 35 of the atlas.
2. Describe the location of the major wheat area using the map on p. 12 of the atlas.
3. Refer to the climate maps on pp. 6-7 of the atlas and find for this region:
   (a) the July temperatures,
   (b) the total yearly precipitation.
4. Climate statistics for Regina, Saskatchewan (574 m in altitude):

	J	F	M	A	M	J	J	A	S	O	N	D
Temp. °C	17	-15	-8	3	11	15	19	18	12	5	-5	-12
Precip. mm	19	17	21	21	40	83	55	49	34	18	20	17

Regina is situated in the major wheat belt. Use the statistics above to complete the following.
   (a) Find the relationship between the highest temperatures and greatest precipitation.
   (b) Calculate the yearly total precipitation and describe it in words.
   (c) Is the July temperature on the low or high side of the range shown on the July temperature map on p. 6 of the atlas?
5. Refer to the soils map on p. 9 of the atlas to name the two soil types in this region.
6. Refer to the map on p. 62 of the atlas to name the landform region in which this wheat farming area is located.
7. What would be the best methods of transportation to get the wheat from Regina to overseas markets? (Refer to the wheat exports map on atlas p. 35.)

Rice Farming

The most important food crop in the world is rice. Rice is the main part of the diet of about 60% of the world's population. It yields more nutrition per acre than any other grain crop, and it has a strong husk which means it stores well.

ACTIVITY 93
1. Refer to the map of world agriculture on atlas pp. 150-151. In which areas of the world is rice the dominant crop?
2. Relate these areas to the world population map on atlas pp. 146-147 to note any relationship between areas where rice is grown and population density.
3. (a) Refer to the economic profiles of Indonesia, India, and Burma on atlas p. 35A. For each country find the percentage that agriculture contributes to the Gross Domestic Product and the percentage of employment in agriculture.
    (b) Write a statement about the relative importance of agriculture in the economies of these countries.
    (c) Compare the percentages in (a) to those for Canada on p. 34A of the atlas.
    (d) How would you describe the importance of agriculture in the economy of Canada compared to its importance in the economies of Indonesia, India, and Burma?
4. Refer to the map of the world: climate regions on atlas p. 144. What name is given to the climate region where most of the world's rice is grown?
5. Refer to the climate graphs of Bombay, Singapore, and Cherrapunji on atlas p. 97.
    (a) Describe the annual temperatures at each of these places.
    (b) Calculate the temperature range for each place.
    (c) What are the total annual amounts of precipitation?
    (d) Describe the annual distributions of precipitation.
    (e) Refer to the rainfall inset map on atlas p. 99. What name is given to the winds that bring the summer rains? On what dates do these winds reach Bombay? Cherrapunji? What does the fact that a

map can be drawn with arrival dates for these winds indicate about their reliability?
- (f) Refer to the world climate map on atlas pp. 142-143. What range of rainfall is prevalent over most of the major rice-growing regions?
- (g) Refer to the map of tropical revolving storms (Northern Hemisphere) on atlas p. 143. What are the names of the storms that can cause great devastation to the rice-growing regions? During what months of the year are they most likely to occur?
- (h) Make a summary statement about the temperature and precipitation conditions necessary for rice cultivation.

6. Refer to the graph of food production and population on atlas p. 150.
   - (a) Compare the percentages of food production and population in the rice-growing areas.
   - (b) Would the bulk of the rice crop be grown for personal use (subsistence), or sale (commercial)?

7. Refer to the graphs of average nutritional levels by continent on atlas p. 150.
   - (a) What does the graph for Asia show about the number of megajoules per day of carbohydrates supplied by the mainly rice diet?
   - (b) Where does Asia stand in relation to the average requirement for good health?

POPULATION

So far we have looked at the physical geography of Canada and other areas of the world. Now it is time to look at the people.

Population Distribution and Density

Two important things to know about the population of a country are where people live (distribution), and how many live in each place (density).

FACTORS THAT AFFECT POPULATION DISTRIBUTION AND DENSITY

1. *Flat Land* — It is easy to construct buildings, roads, and railways on land that is fairly flat. It is also easy to farm flat land, particularly if much machinery is used.
2. *Adequate Precipitation* — People tend to live in areas where there is enough rainfall to grow crops and to feed rivers, lakes, and reservoirs.
3. *Fresh Water Supplies* — We use fresh water from lakes, rivers, and wells every day. Not only do we continually use it in our homes, but many industries require great amounts in the manufacturing process. Agriculture in dry areas or in dry weather may require large quantities for irrigation.
4. *Transportation* — Many lakes and rivers are important transportation routes for moving bulk goods relatively cheaply. The area that has varied forms of transportation is more likely to attract people.
5. *Favourable Temperatures* — People tend not to settle where winters are severe and long, because clothing and fuel cost more. People prefer areas where temperatures are moderate to hot for part or most of the year.
6. *Availability of Power* — Large amounts of different energy sources are required to heat our homes, run our appliances, fuel our cars, and operate our industries.
7. *Resource Availability* — Key mineral resources such as coal and iron ore must be easily available to set up heavy industry.

8. *Good Soil* — People tend to settle in areas where good soil provides a base for varying types of agriculture.
9. *Inertia* — Areas settled in the historic development of a country often remain as population centres. This happens because it is often easier to develop an already settled area than to start a new settlement.

While these are not the only factors affecting where people live, they are the most important. When looking for reasons to explain population distribution and density, check this list to select the most important factors for the area being examined.

ACTIVITY 94
1. Refer to atlas p. 30A to complete the following:
    (a) What is the total population of Canada?
    (b) To what year does this figure apply?
2. Refer to the Canada: population map on atlas p. 10 to answer the following:
    (a) Which areas of Canada are most heavily populated?
    (b) Why do people prefer these areas?
3. What reasons can you suggest for the scattered orange areas and black dots in the northern parts of the provinces and the Arctic?

As you can see from the map on atlas p. 10, the population is not distributed equally among the provinces and territories. This can be seen from this list of the percentages of the total population for each province and territory.

Population of Provinces and Territories: (percent of total), 1981

	%
Alberta	9.2
British Columbia	11.3
Manitoba	4.2
New Brunswick	2.9
Newfoundland	2.3
Northwest Territories & Yukon	0.3
Nova Scotia	3.5
Ontario	35.5
Prince Edward Island	0.5
Québec	26.5
Saskatchewan	4.0
Total:	100%

A better way to see this is to put it into the form of a circle graph. This kind of graph is best for showing how the parts of something make up the whole. To get this information ready for a circle graph the percentages first have to be changed to degrees. This can be done easily as follows:

A circle has 360° — The percentages total 100
1% of the circle will be 360° ÷ 100 = 3.6°

To change the percentages in the table to degrees, multiply each number by 3.6°.

		rounded off
Alberta	9.2 x 3.6 = 33.1	33°
British Columbia	11.3 x 3.6 = 40.7	41°
Manitoba	4.2 x 3.6 = 15.1	15°
New Brunswick	2.9 x 3.6 = 10.4	10°
Newfoundland	2.3 x 3.6 = 8.3	8°
Northwest Territories & Yukon	0.3 x 3.6 = 1.1	1°
Nova Scotia	3.5 x 3.6 = 12.6	13°
Ontario	35.5 x 3.6 = 127.8	128°
Prince Edward Island	0.5 x 3.6 = 1.8	2°
Québec	26.5 x 3.6 = 95.4	95°
Saskatchewan	4.0 x 3.6 = 14.4	14°
	Total:	360°

To draw the graph you need first to draw a circle (the larger the circle the easier it will be to work on) and use a protractor to measure the angles. The circle here has been marked off in 10° units to make it easier to use.

The north line (0°) is used as a starting point for dividing the circle. Each section is then drawn in a clock-wise direction. Before you start, always check to see that the angles add up to 360°. If not, small changes may have to be made to allow for rounding off to the nearest whole number. In some cases it may be best to begin with the largest sectors and work downwards in size. This is not essential. Once drawn, each sector should be identified by a name, shading, or colouring.

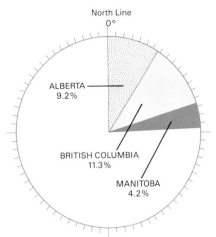

ACTIVITY 95
1. Complete the circle graph.
2. What advantage is there in using a circle graph rather than a list of numbers?
3. Use the factors affecting population distribution to help explain what you see on the circle graph.
4. Each of the regional maps on pp. 20-26 in the atlas has a table giving "Population, 1981." This table gives figures for urban, rural non-farm, and rural farm populations. Construct a circle graph showing this information for your own province.

Population Density

As you have seen from the map and the circle graph, the population is not distributed very evenly across Canada. Neither is the population distributed evenly in each province. This is true for objects as well as people.

ACTIVITY 96
1. Where might you find a high density of the following: cars? books? buildings? wild animals? historic objects?
2. The following are different types of residences: single-family home; low-rise apartment; semi-detached home; high-rise apartment; townhouses. Put them in a list with the one having the highest density of people at the top and the lowest at the bottom.
3. Name two types of buildings in which the density of people will change over a period of time.

4. Name some low-density areas in your own community.

Population density is made up of two variables. These are the number of people, and the size of the area.

$$\text{Population density} = \frac{\text{population}}{\text{area}}$$

Remember that population density presumes that all of the people are spread evenly over an area — as you know this is rarely true.

ACTIVITY 97
1. Each square represents 1 km². The number in each square is the population in that area. Find the total population, the total area, and the population density for each group of squares.

(a)

10	0	2	1
5	3	8	20
0	15	6	26

(b)

0	4	0	3	0	2	1	0
2	0	0	1	5	0	0	0

(c)

62	195	38	63	37	134	125
22	60	0	0	0	105	77
45	68	73	54	100	15	39
320	560	116	403	2312	6	88

156

2. The total population of Canada in 1981 was 24.3 million. The area of Canada is 9.976 million km². What was the population density of Canada in 1981?
3. (a) How many people does each red square on the table of census metropolitan areas on atlas p. 10 stand for?
   (b) Using the squares as a guide, what would be the population of: Halifax, Ottawa, Regina, Vancouver?
   (c) List the cities in this table which have more than 1 million people.
4. Write a statement about where most Canadians live. Compare your statement with the population density of Canada.

ACTIVITY 98
Refer to the population distribution on the following atlas maps: North and South America, pp. 60-61; Europe, p. 79; Asia, p. 101; Australasia, p. 119; Africa, p. 126. Find the following for each continent:
1. (a) The areas of greatest population.
   (b) Suggest reasons for people choosing to live in these areas.
2. (a) The areas of least population.
   (b) Suggest reasons why people avoid these areas.
3. Which one of the maps shows the least amount of lightly populated land?
4. Examine the population density figures for all countries listed on atlas pp. 30A-32A and list the following:
   (a) the highest density and where it is found,
   (b) the lowest density and where it is found.
   (c) List the ten most densely populated countries and the continents on which they are located.
   (d) List the ten least densely populated countries and the continents on which they are located.
   (e) List the ten countries with the greatest populations and the continents on which they are located.
   (f) List the ten countries with the smallest populations and the continents on which they are located.
   (g) Which continent has the greatest number of heavily populated countries?
   (h) Which continent has the greatest number of lightly populated countries?

(i) Use the information from (c) and (e) to make a statement about the relationship between total population and population density.

The following may be useful when describing population densities in words:

almost uninhabited	less than $1/km^2$
lightly populated	$1\text{-}49/km^2$
moderately populated	$50\text{-}99/km^2$
densely populated	over $100/km^2$

PART 4:
Historical Applications

Human Settlement Patterns

In addition to the patterns on the earth's surface created by physical features, the uses people make of the land also create patterns. In many cases these patterns are influenced by the physical environment and the level of technology of the people in the area.

For example, roads and railways — transportation networks — create a pattern in an area. Although the main direction of the road or railway is determined by starting point and destination, some details of the route might be determined by the physical environment. As roads and railways cannot go up and down steep slopes, they may change direction and go a long way around to avoid such problems.

The density of the transportation network — the number of roads and railway lines — reflects the level of technology in the society. In a highly developed society, there may be a number of main highways, many subsidiary roads, and several railway lines within a given area. In a less developed society, only one road may cross a large area, and there would be no railway lines.

Human settlements form very distinctive patterns. There are two main categories of settlement. The first category is dispersed settlements. These vary from widely scattered farmhouses such as are found in the prairie provinces, to the "point form" which is very common in much of northern Canada. Residential centres in the north are located hundreds of kilometres from one another and they may or may not be linked by road.

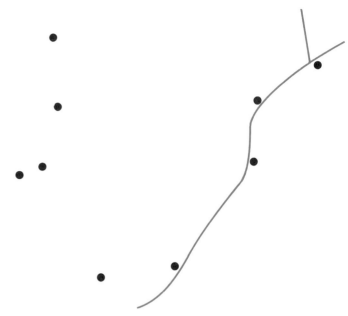

The second category of settlement is called compact or nucleated. In this type the buildings are clustered together. Compact settlements create varied patterns. In the linear form most of the buildings are arranged along one street or road. Examples can be found along railway lines and highways in Central Ontario, in the valleys of British Columbia, and along the north shore of the lower St. Lawrence River.

Rectangular forms are found in settlements where streets cross at right angles.

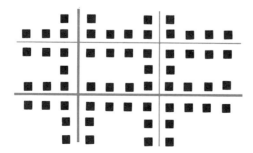

An early version of the compact settlement was the round village, where the houses were set out in a radial pattern. The round village was often protected by a wall, and as the settlement grew, new walls were constructed farther from the centre of the village.

This cooperative town of Nahalal, Israel, is arranged in a radial pattern. The business district is at the centre, surrounded by residences. These in turn are surrounded by cultivated fields.

The type of settlement may be determined by the physical environment. In Japan, for example, there is a large population and a small area of farmland. The houses in Japanese farming villages are often tightly packed together because of the need to use as much land as possible for farming.

REASONS FOR THE LOCATION OF SETTLEMENTS

There are usually specific reasons why a settlement has been located in a particular place. In some cases the physical environment is a factor: in deep mountain valleys north-facing slopes receive only a few hours of sunshine, even in summer. Settlements would therefore tend to be located on the south-facing slopes.

Another reason is a combination of physical factors and the historic situation when the settlement was started. Before there were highways, rivers provided the best means of transportation. The most common early locations for settlements were along rivers, or at the confluence of two rivers which gave an even better access to a large territory.

Another useful location was at the ford of a stream where eventually a bridge or ferry service would be developed. Later, when roads had been built, settlements started at or near the intersection of two or more roads to facilitate trade and communication. In hilly or mountainous land, a settlement might have been located on a hilltop for reasons of defence. All settlements, of course, had to be near resources such as water, wood, and productive farmland.

Once a settlement has been established, inertia will cause it to continue in that location even though the original reason for its location no longer applies. Many settlements that were originally established for one purpose — e.g. defence or trade — probably will have developed multi-purpose functions over a long period of time.

Settlement in much of the Old World was affected by the factors described above.

In the New World, and particularly in many areas of Canada, the land was surveyed before settlement began. The plan of land settlement may have disregarded some of the physical or cultural factors described above.

TOWNSHIPS IN THE PRAIRIE PROVINCES

The Prairies is an area where the land was surveyed before settlement. The township size selected for the prairie provinces was similar to that of the American township: square with sides 9.6 km long, divided into thirty-six sections of 2.59 km^2. Each section was then subdivided into four homesteads of 64.7 ha or into sixteen local subdivisions of 16.2 ha. The system was later modified slightly, but the general township pattern is still evident in the prairies today.

Settlement Growth

Many of the present settlements in Europe have grown up over a long period of time. This activity will show you how this growth has taken place.

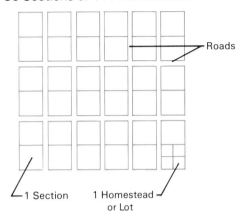

PRAIRIE TOWNSHIP
36 Sections or 144 Homesteads

1 Section   1 Homestead or Lot

ACTIVITY 99

The map on p. 166 shows an area in which an early tribe wishes to settle. Most of the area is covered by forest except for the hilltops and some marsh land. The marsh consists of low, wet land covered with tall reeds and scattered patches of small shrubs and trees.

Place a piece of tracing paper over the map, and trace all the features shown on the map, including the grid and its numbers. Or obtain grid paper with squares as large as those on the map, and copy all the map — the grid numbers and letters will assist you in copying all features accurately. Lightly shade the hilly areas as on the map.

You are going to show the development of settlement over time on this map, according to the following instructions.

Use different colours for each time period and capital letters to indicate the type of buildings, e.g. W-warehouse, F-factory, H-housing. Make a key for every colour and symbol used. You may develop a map individually or in groups.

Period 1
The basic needs of the tribe include a constant supply of fresh water for cooking and drinking, arable land for farming, grazing land for cattle, a regular supply of fuel for cooking and warmth in winter, and building materials for homes. The tribe needs a site that is easily defended against other tribes that live in the area.
   (a) Choose a site for your village that best meets all the needs of the tribe. Mark the site with a dot.
   (b) Mark two sites which have been settled previously by other tribes with the same needs.
   (c) What materials would be used for building homes?
   (d) Justify your choices of the three sites.

Period 2 (use a different colour)
   (a) Many years have passed. Roads now connect the original 3 settlements and one of them has become a market town. (Mark in two whole squares for this town.)
   (b) The marsh has been drained by a canal (mark it in), a ferry crosses the river (mark it in), and a bridge is built over the river and the canal (mark it in).
   (c) Two new roads cross the whole area, meeting at the bridge (mark them in).
   (d) Boat traffic can come up the river as far as the bridge. The goods they bring in and pick up are stored in warehouses along the river (mark in 2 squares for warehouses).

Period 3 (use a different colour)
   (a) After many more years, the town has increased in size (mark in 4 more squares for housing and 2 more for warehouses).
   (b) A new sawmill is built to handle logs coming down the river (mark in 1 square), and a woollen mill is constructed to handle the wool obtained from sheep which graze on the hillside (mark in 1 square near the river).
   (c) Mark in 2 more squares for housing near the sawmill and/or woollen mill.

Period 4 (use a different colour)
   (a) Coal is found in the hills in the northeast part of the map (mark in 6 squares). A mining town develops near the coal mine (2 squares), and the newly developed railway technology is used to build a railway link joining the mining town and the port (mark it in).
   (b) Coal production and shipment increase traffic greatly on the river. New docking facilities are established (2 squares) because larger ships cannot travel upriver, and housing (4 squares) is built to accommodate dock workers.
   (c) Business flourishes and additional warehousing (4 squares) is established near the docks.
   (d) Factories (4 squares) are built to help meet the needs of a growing population, and it soon becomes necessary to build a north-south railway (mark it in) and a major road through the area which passes through the city (mark it in).
   (e) More housing (5 squares) is required and it is built away from the factories and port. As the economy grows, people spend part of the year at a small resort town (2 squares) on the coast. It is linked to the port by a railroad (mark it in).

Period 5 (use a different colour)
   (a) Automobiles are widespread. People are able to live outside the city and commute back and forth to work. Two suburban towns are built outside the city (2 squares each).
   (b) A large dock is built at the river mouth to handle deep water traffic (3 squares). Housing (2 squares), associated with the dock development, is constructed.

Period 6 (use a different colour)
   (a) The settlement has grown (8 new squares for housing). Traffic has increased greatly, so a major highway is built through the area, passing near the city (mark it in).
   (b) Three more commuter suburbs (2 squares each), are established outside the city.

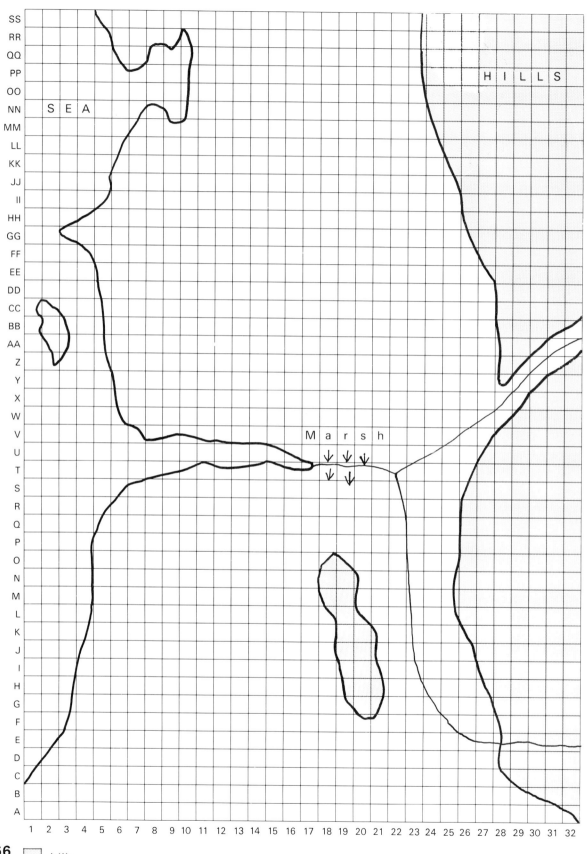

166 hills

(c) Factories (4 squares) are now established near the major highway which provides easy access to other areas.
(d) Environmental concerns arise so a natural park area is established in the hills (4 squares).

Summary
1. Name the main factors that influenced your decisions about locating places for each time period.
2. List the significant methods of transportation for each time period and explain how each affected urban growth.
3. Compare your map to others in the class. How are they similar and different? Why are they not all the same? Is any one map more correct than another? Why?
4. If you could redraw your map, what planning would you use to make this area a more pleasant place in which to live?
5. Refer to the urban land use maps of Paris and Moscow on p. 92 of the atlas:
   (a) What relationships can you find between land use and rivers, land use and roads, land use and railways?
   (b) What land uses would you associate with the historical core of each city?
   (c) Do any of the decisions you made about the location of specific land uses seem to be reflected in either of these maps? Which decisions were they?

ACTIVITY 100

Locate Halifax on the map of Eastern North America on atlas pp. 50-51.
1. (a) In what part of Canada is Halifax located?
   (b) In what province is Halifax located?
   (c) Describe the location of Halifax within the province.
   (d) Describe the location of Halifax in relation to the Atlantic Ocean and Europe.
2. In 1749 the British sent colonists to settle Halifax as a counterbalance to the French fortress of Louisbourg, so that ships sailing out of Boston would be given protection by the military and naval base at Halifax.
   (a) Give the position of Halifax in relation to Louisbourg and to Boston.
   (b) Use the map scale to find the distance from Halifax to Louisbourg and from Halifax to Boston.
3. Refer to the map on atlas p. 62.
   (a) In what landform region is Halifax located?
   (b) Describe the appearance of this region.
   (c) Where is farming possible within this type of region?
4. Refer to the maps of climate and climate regions on atlas pp. 6-8, and the climate graph for Halifax on atlas p. 7.
   (a) What are the average January and July temperatures in Halifax?
   (b) Calculate and describe the annual temperature range.
   (c) What is the length of the growing season?
   (d) What is the exact altitude of Halifax?
   (e) What climate controls have the greatest effect on temperature?
5. (a) What is the average January and July precipitation in Halifax?
   (b) What is the total annual precipitation?
   (c) How much snowfall does Halifax get?
   (d) Describe the yearly distribution of precipitation.
   (e) What controls have the greatest influence on precipitation in this area?
   (f) Refer to the map of climate regions on atlas p. 8. Name and describe the region in which Halifax is located.
   (g) How does the climate affect the farming in the Halifax area?

6. Halifax is situated on a natural harbour protected by a 20 km inlet. In its early years it was an important part of the triangular trade pattern that existed between Britain, the Caribbean, and North America.
   (a) Refer to the map of the Atlantic Ocean on atlas p. 131. What ocean currents affected ships sailing from Britain to the Caribbean, the Caribbean to Halifax, and Halifax to Britain?
   (b) Refer to the maps of January and July winds on atlas pp. 142-143. What two major wind systems influenced this trade pattern?
   (c) What information on atlas p. 131 represents a limitation of the coastal location of Halifax?
7. Refer to the map of vegetation on atlas p. 9.
   (a) In what vegetation zone is Halifax located?
   (b) What were the settlers' main uses for the trees?
8. (a) Why did people originally settle in the site of Halifax?
   (b) What difficulties caused by environmental factors did these early settlers face?
   (c) What were the natural advantages of this site?
   (d) What industry was important in Halifax during the period of trade described above?
   (e) What resources in the natural environment were useful for this industry?

TRAVEL

Before the development of the steam engine, travel on land was on foot, on horseback, or in a vehicle drawn by horses. Inland waterways — rivers and canals — were also used for transportation both before and after the arrival of the railway. Travel at sea was by sailing ship.

ACTIVITY 101
1. (a) Early travellers had to travel by rivers. Which river pattern would have been most useful for exploring large areas? (Refer to p. 108 in this book.)
    (b) Name three European rivers that have this pattern.
Refer to the map of Southern Europe and North Africa on atlas pp. 90-91.
2. What are the two most common factors that account for the location of most of the cities marked on this map?
3. Use the map scale to find the shortest distances between the Rhine River system and the Danube river system; the Vistula River system and the Dneiper River system.
4. (a) Approximately how long a journey would be made by early traders travelling from the mouth of the Danube River to the mouth of the Rhine River overland?
    (b) Give the latitude and longitude of the beginning of the branch of the Danube River near Basel.
    (c) How long would the journey in (a) be, if taken by sea?
5. The Plague ("Black Death") was brought to Europe in ships from the Near East.
    Which areas of Europe would be the first to be infected? the last?

ACTIVITY 102

Method of Travel	Speed in km/h
(a) walking	6.0
(b) canoe	5.0
(c) horseback (walking pace)	7.0
(d) train	80.0
(e) car	96.0
(f) plane	800.0

1. Refer to the map of China on atlas pp. 112-113 and estimate the distance along the Huang River from its mouth (ignore the delta) to the city of Xi'an.
2. (a) Use a piece of paper to mark off the actual length and find the approximate distance by checking with the linear scale.
   (b) What was the difference between your estimated distance and the actual distance?
3. Use the actual distance to calculate the time required to travel that distance by each method of travel listed.
4. Construct a graph to show the total length of time required to travel the river by each of the methods in the table.
5. Use the colour key for altitude as a guide to describe whether the journey was easy or whether you might have encountered any difficulties with any of the methods of travel.

ACTIVITY 103

Refer to the map of Southern Europe and North Africa on atlas pp. 90-91.
1. List the factors that made travel within Europe and between Europe and Asia and Europe and Africa fairly easy.
2. What factors would hinder travel in this region?
3. Assume the role of leader of one of the following:
   (a) some Huns moving into Europe from the north coast of the Black Sea
   (b) some Crusaders travelling from Vienna to Constantinople
   (c) some Swedish Vikings going to Constantinople.

As leader of the group, describe the directions you would take; name the main rivers you would use; find the highest and lowest land elevations you would encounter; calculate the distances you would travel.

ACTIVITY 104

Read the following description of a voyage made by a fleet of Spanish ships on their way from the Caribbean to Spain in 1622.

> The long, restless night had ended. Dawn awoke the crew and passengers to gale force winds shrieking in from the northeast. As the winds crossed the Gulf Stream current, the confused seas towered into huge white-capped waves that tossed the ships to and fro until even the experienced sailors were overcome by seasickness. At first the fleet of 28 vessels bound for Spain, carrying the riches of the New World, was able to keep its sailing order. But, as the day wore on, the winds rose to hurricane force and the ships lost sight of each other in the towering waves. The *Santa Margarita*, with a broken mainmast and splintered tiller, was swept along by the huge waves during a night of fear and terror. A wind shift at midnight drove the treasure-laden galleon northwards towards the reefs and shoals of the Florida Keys. Sunrise the next morning saw the little vessel being pushed inexorably towards the great rolling combers foaming over the reef. Suddenly the *Santa Margarita* struck the shoal and the powerful waves began to break the vessel apart. By the time the storm ended 8 ships of Spain's Tierra Firme fleet had been sent to the bottom and the cargo vital to the economy of Spain would never reach home.
>
> based on the article
> "Treasure From the Ghost Galleon: *Santa Margarita*,"
> by Eugene Lyon, *National Geographic Magazine*,
> February 1982

Refer to the map of winds, ocean currents, and tropical revolving storms on pp. 142-143 of the atlas.
1. What winds and ocean currents would be used by the Spanish seamen to go from Spain to the Caribbean?
2. What winds and ocean currents would be used to return to Spain from the Caribbean?
3. What was the name of the type of storm that sank the *Santa Margarita*? What are these storms called in other parts of the world?

4. At what time of the year was the fleet returning to Spain?
5. What conditions accounted for so many shipwrecks in this area between Cuba and Florida? (Refer to the map of the Caribbean on atlas pp. 74-75.)

ACTIVITY 105
Refer to the climate graphs of Havana and Mexico City on p. 55 of the atlas and the illustration of the Spanish soldier.
1. What problems would exist for soldiers wearing this type of armour in the Caribbean area?
2. What adaptations in life style would have to be made by the explorers in this area in order to carry out their tasks efficiently?

ACTIVITY 106
1. Which of the major river systems would have been most useful to explorers in large parts of North America before any road or rail transportation existed? Justify your answer.
2. Refer to the map of Eastern North America on atlas pp. 50-51.
    (a) Locate the following water bodies which were important for the fur trade:

    St. Lawrence River   Lake Nipissing
    Richelieu River   Georgian Bay
    Lake Champlain   Lake Ontario
    Hudson River   Mohawk River
    Ottawa River   French River

    (b) Calculate the length of the St. Lawrence River from Montreal to where the river flows into the Gulf of St. Lawrence. Calculate the width of the river at its widest point.
    (c) In what direction does the St. Lawrence River flow? Why was this a disadvantage for the fur trade?
    (d) You are sailing with Jacques Cartier up the St. Lawrence River, and you are the official geographer to the expedition. Describe each of the following in your log book: shape of the land, vegetation, course of the river, wildlife, and Indian settlements. Draw a profile of the river and shoreline to help explain your written description.

(e) What main rivers, vital to the fur trade, were controlled by the Iroquois?
(f) What effect did this have on the fur trade and the alliances formed between Indians and Europeans during that period?
(g) Explain the reasons for the success of the settlement at Québec and the failure of the settlement at Port Royal.
3. Describe each settlement on the map on atlas pp. 50-51 in relation to its access to water.

Following the defeat of the Iroquois, the French were able to move into the region south of the Great Lakes. In order to do this they had to cross a watershed, or height of land. The topography of the land enabled them to accomplish this quite easily.

The spillways formed during the period of glaciation made excellent portages for the French travelling south, since they provided relatively low passes through the surrounding hills. And not far away were the headwaters of streams which flowed into lakes and rivers of different systems. The explorers discovered these pathways, through which they portaged from the Great Lakes to the Mississippi River system.

ACTIVITY 107
1. What early methods of transportation made the beaver an easy prey for trappers?
2. In what ways did glaciation produce a suitable habitat for the beaver?
3. Refer to the world climate maps on pp. 142–143 of the atlas. What climatic conditions made exploration in the Mississippi drainage basin more comfortable than in the area of the Great Lakes?

ACTIVITY 108
1. (a) Draw a climate graph for Brest, France using the statistics given below:

Location: 48° 19'N 04° 47'W
Altitude: 17 m

	J	F	M	A	M	J	J	A	S	O	N	D
Temp. °C	7	7	8	10	13	16	17	18	16	13	9	8
Precip. mm	89	76	64	64	48	51	51	56	58	91	107	112

  (b) Interpret the graph using the questions on pp. 136-137 in this book.
2. (a) Interpret the graph of Québec on atlas p. 7.
  (b) Interpret the graph using the questions on pp. 136-137 in this book.
3. Compare these two locations as follows:
  (a) what are the latitudes of each?
  (b) which location has the more extreme temperatures?
  (c) which location has the longer growing season?
  (d) account for the differences in temperature;
  (e) what adaptations in life style would have to be made by people moving from Brest to participate in the fur trade or establish farms in the Québec City region?

ACTIVITY 109
1. European explorers introduced the horse to the North American continent. Refer to the maps of North America climate and North America vegetation on atlas pp. 56 and 63.
    (a) What climate and vegetation conditions favoured the rapid spread of horses in North America?
    (b) What effects did the introduction of horses have on the Native Peoples?
2. In summary, list the major climatic factors that had a significant effect on the exploration and settlement of North America.

GLOSSARY

Alluvium: silt, sand or gravel deposited by a stream
Altitude: height above mean sea level
Ancient Shield: a very old mass of rocks that has remained fairly stable since an early period in the earth's history
Antarctic Circle: the 66.5°S parallel of latitude
Arable and Permanent Pasture: a land-use classification designating grazing areas and land suitable for farming
Arable Land: farmland which is ploughed and cultivated or which is suitable for growing crops
Atlas: a figure in Greek mythology who held up the sky with his head and hands; a collection of maps
Atmosphere: the gaseous envelope surrounding the earth
Arctic Circle: the 66.5°N parallel of latitude

Background: the part of a photograph farthest away from the viewer and showing the least detail
Bedrock: solid rock underlying the loose soil on the surface of the earth
Broadleaf Forest: a forest area in which most of the trees are of the deciduous variety with wide leaves
Build: the basic structures which form the continents

Carbohydrate: an energy-producing organic compound of carbon with oxygen and hydrogen, e.g. starch, sugar
Champlain: Samuel Champlain, a French explorer actively involved in attempting to establish a colony in New France in the early 1600s
Climate: average weather conditions over a broad area for a period of many years
Climate Graph: a graph showing monthly temperature and precipitation conditions at a particular location
Cold Front: the boundary zone between a mass of warm air and an advancing wedge of cold air

Compact Settlement: a settlement in which buildings are clustered together
Compass Bearing: the angular difference between magnetic north and a point on the earth's surface. It is measured in degrees clockwise from magnetic north.
Confluence: the point at which a tributary joins the main stream
Conformal: the name given to a map projection in which shape is maintained over a small area
Conical Orthomorphic Projection: a projection based upon a cone which cuts the globe at two standard parallels. The word orthomorphic means true shape.
Coniferous Forest: a type of forest in which most of the trees are evergreen, cone-bearing with needle-leaves
Continent: a major land mass of the earth
Continental Climate: the climate of an interior area of a continent, which has seasonal extremes of temperature and low precipitation occurring mainly in the summer
Convectional Precipitation: precipitation resulting when warm air from a heated land surface rises and is cooled to form clouds from which heavy precipitation falls
Cross Section: a drawing which gives a cut-away side view of the land at and below the surface
Cyclonic Precipitation: precipitation which occurs along fronts caused by air masses moving over or under each other
Cylindrical Projection: a projection based on the idea of a cylinder surrounding the globe and touching it along the equator or pair of central meridians, or cutting the surface at parallels 45°N and 45°S

Delta: a land area built of material deposited at the mouth of a river as it slows down upon meeting the sea
Dendritic Drainage Pattern: a tree-like pattern of tributaries converging on a main stream

Direction: a point to or from which a person or thing moves

Dispersed Settlement Pattern: a rural settlement pattern of isolated farms or buildings

Distribution: the way things are spread over the land surface

Divide: an area of high ground between river basins

Drainage Basin: an area drained by a single river system

Dune: a ridge of drifted sand

Eastings: the distance east from the origin of a grid

Elevation: the height above some particular level

Equator: the 0° parallel of latitude midway between the Poles

Equatorial Scale: distance measured along the equator; on some map projections the scale is only accurate a few degrees north or south of the equator

Foreground: the part of a photograph nearest the viewer and showing the most detail

Frontal Precipitation: precipitation occurring along a front when two air masses meet (see Cyclonic Precipitation)

Geology: the science that studies the history, composition, structure, and processes of the earth

Glacier: a mass of ice moving slowly outward from an area of snow accumulation

Gneiss: a coarse-grained crystalline rock with a banded appearance

Granite: a coarse-grained igneous rock consisting mainly of quartz, orthoclase, felspar, and mica

Great Circle: any circle on the earth's surface which cuts the globe in two equal parts. It is the shortest distance between any two points on the earth's surface.

Greenwich Meridian: the Prime Meridian which passes through the Royal Observatory at Greenwich

Grid: a network of squares formed by lines drawn parallel and at right angles to a central axis from which the position of any place can be stated

Gross Domestic Product: the value of goods and services produced in a country before providing for depreciation or capital consumption

Ground-level Photograph: a photograph taken by a camera held by a person standing on the ground

Growing Season: the part of the year when temperatures are high enough to allow plant growth

Headwaters: the upper parts of a river system including its source streams

Height of Land: an area of high ground separating major drainage basins (see Divide and Watershed)

Hemisphere: half a sphere. The earth is divided into the Northern and Southern Hemispheres by the Equator and the Eastern and Western Hemispheres by the Prime Meridian and 180° meridian

High Latitudes: the areas of the world between 60° north or south of the Poles

Hills: uplands with lower elevations than mountains in the same area

Horizon: the boundary where earth or sea and sky appear to meet

Horizontal Axis: a horizontal line used as a point of reference on a graph

Hot Desert: a desert area with high daytime temperatures most of the year

Humus: the remains of decomposing plants in the soil

Igneous Rock: a rock which has been formed by the solidification of molten material

Interior Plains: a physical region of low local relief in western Canada between the Canadian Shield on the east and the Cordillera on the west

International Date Line: a line following approximately the 180° meridian used in the time zone system to indicate the beginning of a new day and the end of the previous day

Island: a piece of land surrounded by water

Isohyet: a line joining places having equal amounts of precipitation

Isotherm: a line joining places having equal temperatures

Land Depression: an area of land with an elevation below sea level

Land Use: the uses made of the land by people

Latitude: the angular distance of any point north or south of the Equator and measured in either direction to 90°

Layer Tinting: a method of using different colours on maps to indicate various ranges of altitude

Leeward: the side away from, or protected from, the wind

Legend: an explanation of the symbols, shading, or colours used on a map

Limestone: a rock formed under ocean water and consisting mainly of calcium carbonate

Line Scale: a line divided into units and used for measuring distance on maps

Linear Settlement: a long drawn-out settlement formed by the clustering of buildings along a major route

Longitude: the angular distance of any point east or west of the Prime Meridian and measured in either direction to 180°

Lowlands: areas of land at low altitudes

Low Latitudes: the areas of the world between 30° north or south of the Equator

Loxodrome: a line of constant compass direction which forms a straight line on a Mercator projection

Map Projection: a representation of the earth's surface on a flat map

Marine Climate: a climate experienced mainly on mid-latitude west coasts, usually with small daily and seasonal temperature ranges and appreciable cloud cover and precipitation

Mean Annual Precipitation: the average amount of precipitation at a given location in one year

Megajoule: a unit of nutrition measuring the energy produced by the body as it "burns" food

Mercator: Gerardus Mercator, a Flemish geographer who lived during the 1500s

Mercator Projection: a cylindrical projection used by Mercator for his world map of 1569

Meridians: lines of longitude joining the two poles and forming half of a great circle

Metamorphic Rocks: rocks changed from their original form into other forms by great heat and pressure

Mid Latitudes: the areas of the world between 30° and 60° north and south of the Equator

Modified Gall Projection: a cylindrical projection which is a reasonable compromise between accuracy of shape and area

Moraine: a mass of clay and stones carried and deposited by a glacier and often in the form of low hills

North Pole: the geographic point at which all meridians meet in the Northern Hemisphere

Northings: the distance north from the origin of a grid

Nucleated Settlement: a type of settlement in which the buildings are clustered together

Oblique Air Photograph: a photograph taken from the air with the camera pointing down at an angle

Oblique Mercator Projection: a type of map projection based on the idea of a cylinder touching the globe along lines other than the equator or pair of meridians

Ocean Currents: surface currents set in motion by prevailing winds

Open Woodland: a forested area with widely-spaced trees

Orient: to set a map so that its north is parallel to north on the earth's surface

Orographic Precipitation: precipitation caused by the rising of moisture-laden air over a mountain range

Outline Map: a map showing the outline of an area of the world with no detail filled in

Parallels: lines of latitude joining all places with the same angular distance north or south of the Equator and parallel to it

Pasture Land: land used for the grazing of livestock

Peninsula: a piece of land projecting into a sea or lake

Physical Map: a map which shows the shape and form of the land surface

Plateau: an upland with level ground

Plate Tectonics: the theory that explains the origin of the structural features of the earth in relation to the movement of the earth's plates

Political Map: a map which shows the countries or other political divisions of the world

Population Density: the average number of people within a given area

Population Distribution: where people live within a given area

Prairie: an extensive area of treeless grassland

Precipitation: all forms of deposition from the atmosphere including rain, snow, sleet, hail, and dew

Profile: an outline of an object

Radial Settlement Pattern: a pattern of settlement in which the streets and houses spread out from a central point in ever-increasing circles

Radial Stream Pattern: a pattern of streams flowing outward from a dome or cone-shaped upland

Rain Shadow: the drier, leeward side of a mountain area

Rank Order: to arrange numerical data in descending or ascending order based on the specific topic under consideration

Recent Deposits: loose materials deposited on the surface of the earth by wind to form dunes or by water to form deltas

Rectangular Settlement Pattern: a pattern of settlement formed where streets cross at right angles

Region: an area of the earth's surface defined by particular characteristics

Regional Topographic Map: larger scale maps in an atlas which show the shape and form of the land surface

Relief: differences in altitude of the earth's surface

Representative Fraction: the ratio which the distance on a map bears to the real distance on the earth's surface

Rhumb Line: a line of constant compass direction (see Loxodrome)

Rural: the country as opposed to the city or urban area

Rural Non-farm: a census classification designating people who live in the country but do not operate farms

Sandstone: a sedimentary rock formed by the cementing together of grains of sand

Scale: the ratio between distance on a map and the real distance on the earth's surface

Schist: a metamorphic rock with a wavy texture

Sediment: particles of sand or clay laid down by rivers, wind, glaciers, and seas

Sedimentary Rocks: rocks made up of sediments laid down in layers and cemented together

Shale: a sedimentary rock formed by the cementing together of fine mud deposits

Soil: the thin surface layer of the earth made up of mineral particles, decaying vegetation, and living organisms

South Pole: the geographic point at which all meridians meet in the Southern Hemisphere

Spillway: a valley created by a stream that had drained a lake during a former period of high water

Spot Height: a precise measurement of altitude shown on a map or marked on the ground by a surveyor

Statement Scale: a map scale expressed as a ratio or a statement

Subsistence: a form of agriculture in which all crops grown are consumed by the farmer and his or her family

Symbols: marks, lines, or colours used on maps to represent real things on the earth's surface

Thematic Map: a map of an area of the world dealing with a specific theme or topic

Topsoil: cultivated soil

Transverse Mercator Projection: a form of the Mercator projection based on the idea of a cylinder surrounding the globe and touching it along a pair of central meridians

Trellis Pattern: a rectangular drainage pattern found in hill areas composed of alternate hard and soft rocks

Tributary: a stream or river which joins a larger one

Tropical Rainforest: a vegetation classification designating areas that are heavily forested as a result of high annual temperatures and heavy annual precipitation

Tropic of Cancer: the 23.5°N parallel of latitude

Tropic of Capricorn: the 23.5°S parallel of latitude

Tundra: the vast level Arctic region between the northern limit of trees and the area of permanent snow and ice

Uplifted Mountains: worn-down, old mountain systems that have been lifted higher again by pressures within the earth

Vertical Air Photograph: a photograph taken from the air with the camera pointing directly at the ground

Vertical Axis: a vertical line used as a point of reference on a graph

Warm Front: a boundary zone where a mass of warm air is rising above the cold air which it is overtaking

Watershed: the line separating streams flowing into different drainage basins (see Divide and Height of Land)

Weather: the day-to-day nature of the atmosphere

Wind: air in horizontal motion

Windward: the side facing the wind

Younger Fold Mountains: relatively young mountains which formed when the plates making up the earth's surface moved slowly against each other causing the crust to buckle

Zenithal Equal Area Projection: a form of map projection which provides true equal area

Zenithal Equidistant Projection: a form of map projection in which both direction and distance from the centre of the map are correct

INDEX

A
AGRICULTURE, 146-151
   (see also farming)
   factors that affect farming, 146-148
   wheat farming, 149
   rice farming, 150
AIR MASS, 125-126
AIR PHOTOGRAPHS, 89-94
   oblique, 89-91
   vertical, 92-94
AIR PRESSURE, 116
ANCIENT SHIELDS, 101-102, 106
ANTARCTIC CIRCLE, 36, 37
ARABLE LAND, 146
ARCTIC CIRCLE, 36, 37
ATLAS
   character in Greek mythology, 1
   concept of, 1-3
   scale, 60-65
   types of maps, 66-71
      physical, 69
      place name or political, 66-67
      regional topographic, 70-71
      thematic, 68

B
BACKGROUND, 88
BAR GRAPHS, 132-135
BOUNDARIES, 97, 141
BROADLEAF FOREST, 141, 143

C
CANADA
   agriculture, 146-149
   boundaries, 97
   cross section, 100
   exact position, 96
   location, 95
   population, 152-157
   profile, 99
   shape outlines — provinces, territories, islands, 9-10
   size, 97
   time zones, 54
CARTOGRAPHER, 74
CHAMPLAIN, 2

CHARTS
   kilometric distance, 6
CIRCLE GRAPH, 154-155
CLIMATE, 113-140
   (see also precipitation; temperature)
   cold front, 126
   controls — precipitation, 124-126
   controls — temperature, 118-122
   convectional precipitation, 125
   cyclonic precipitation, 126
   graphs, interpretation, 136-137
   graphs, precipitation, 132-135
   graphs, temperature, 129-131
   growing season, 131
   isotherm, 113-114
   isohyet, 123
   orographic precipitation, 124
   precipitation maps, 123-124
   precipitation range, 123
   regions, 140
   temperature maps, 113-115
   temperature range, 115
   warm front, 126
   word description, 137
CLIMATE CONTROLS
   Edmonton, 138-139
   precipitation, 124-127
   temperature, 118-122
CLIMATE GRAPHS, 128-137
   interpretation, 136-137
CLIMATE REGIONS, 140
COLD FRONT, 126
COLOUR ON ATLAS MAPS, 24-26
COMPACT SETTLEMENT PATTERN, 160
COMPASS BEARINGS, 84-86
  on a mercator map, 85-86
CONICAL ORTHOMORPHIC PROJECTION, 76
CONIFEROUS FOREST, 141, 143
CONVECTIONAL PRECIPITATION, 125
CONVENTIONAL SYMBOLS, 28
CROSS SECTIONS, 100
CYCLONIC PRECIPITATION, 126-127
CYLINDRICAL MAP PROJECTIONS, 77-79

D
DENDRITIC DRAINAGE PATTERN, 108

DENSITY OF POPULATION, 152-153, 155-158
DIRECTION, 30-31, 43-45
DISPERSED SETTLEMENT PATTERN, 159
DISTRIBUTION, 22-23
    planned, 22
    population, 152-157
    unplanned, 22
DIVIDE, 109
DOLDRUMS, 116
DRAINAGE BASINS, 109-111

E
EARTH GRID, 32, 42-45
EARTHQUAKES, 104
EASTERN HEMISPHERE, 40
EASTINGS, 20
EQUATOR, 33-34, 36
    as great circle, 83

F
FARMING, 146-151
    factors affecting:
        climate, 146-147
        landforms, 147-148
        markets, 148
        natural hazards, 148
        soils, 148
    rice farming, 150-151
    wheat farming, 149
FOLD MOUNTAINS, 104
FOREGROUND, 88
FRONTAL PRECIPITATION, 126-127
    (see cyclonic precipitation)
FROST FREE SEASON, 147

G
GAZETTEER, 46-47
    of Canada, 46
    of World, 46
GLACIATION, 110-112
GRAPHS
    bar (precipitation), 132-135
    circle, 154-155
    climate, 128-137
    line (temperature), 129-131
        northern hemisphere, 130
        southern hemisphere, 130
        tropics, 130

GREAT CIRCLE ROUTES, 82-83, 86
GREENWICH MERIDIAN, 39, 52
GRIDS, 15-21, 93
    battleship, 16
    earth grid, 32, 42-45
    eastings, 20
    northings, 20
    number-letter, 15-17
    number-number, 18-21
    plotting a point, 20-21
    plotting position, 18-19
GROUND LEVEL PHOTOGRAPHS, 87-88

H
HEADWATERS, 109
HEIGHT OF LAND, 109
HEMISPHERE
    eastern, 40, 42
    northern, 33-35
    southern, 33-35
    western, 40
HIGH LATITUDES, 38
HIGH OBLIQUE PHOTOGRAPH, 89-91
HIGH PRESSURE SYSTEMS, 116
HILLS, 101
HUMAN SETTLEMENT PATTERNS, 159-162
    compact, 160
    dispersed, 159
    nucleated, 160
    point form, 159
    reasons for the location of, 162
    rectangular, 160
    round, 161
    settlement growth, 163-167
    townships in the Prairie Provinces, 163

I
INTERNATIONAL DATE LINE, 52
INTERSECTION, 20
INTERTROPICAL CONVERGENCE ZONE, 116
ISOBAR, 114
ISOBATH, 114
ISOHYET, 114, 123, 127
ISOSEISMIC, 114
ISOTHERM, 113-114

K
KILOMETRIC DISTANCES CHART, 6

L

LAND DEPRESSIONS, 24, 25
LANDFORMS, 101-107
   ancient shields, 101, 106
   recent deposits, 105
   sedimentary rocks, 102-103, 107
   uplifted remains of ancient mountain systems, 103, 106
   younger fold mountains, 104-105, 107
LATITUDE, 32-38
   distance between parallels, 34
   high, mid, and low, 38
   parallels, 32, 33-35
   parallels show direction, 43
   range, 96
   special parallels, 36-37
LAYER TINTING, 24-25
LEGEND, 4-5
   road map, 5
LINE GRAPHS, 129-131
LINE SCALE, 56-57
LONGITUDE, 39-43
   definition, 39
   Greenwich (Prime) Meridian, 39, 41
   International Date Line, 52
   meridians, 39-41
   meridians as great circles, 83
   meridians show direction, 43
   range, 96
   time zones, 48-52
LOW LATITUDES, 38
LOW OBLIQUE PHOTOGRAPH, 89-91
LOW PRESSURE STORMS, 126
   (see cyclonic precipitation)
LOW PRESSURE SYSTEMS, 116
LOXODROME (RHUMB LINE), 85-86

M

MAP
   Atlas, 1
   atlases, 1,3
   Australia, 13
   Canada's time zones, 54
   Champlain's 2
   outline, 8-11
      islands of Canada, 9
      provinces of Canada, 10
      continents, 11
   Ptolemy's, 1
   world's oldest, 2

MAP BASICS, 4-7
   legend, 4-5, 28
   scale, 6-7
   title, 4
MAP PROJECTIONS, 72-81
   conical orthomorphic, 76
   cylindrical, 77-79
   mercator, 77
   modified Gall, 79
   oblique mercator, 78
   transverse mercator, 78
   zenithal equal area, 81
   zenithal equidistant, 80
MAP SCALE, 56-61
   line scale, 56-57
   representative fraction, 57
   road map, 6-7
   atlas map, 60-65
   statement scale, 56
MAP SYMBOLS, 12-14
   atlas map, 27-29
   conventional, 28
   line, 27-28
   mark, 27
MARINE CLIMATE, 121
MEAN ANNUAL PRECIPITATION, 123
MERCATOR, 1
   compass bearings on a mercator map, 85-86
   mercator projection, 74, 77-78
   oblique mercator projection, 78
   transverse mercator projection, 78
MERIDIANS OF LONGITUDE, 39-43
   (see longitude)
MID LATITUDES, 38
MODIFIED GALL PROJECTION, 79
MORAINES, 109
MOUNTAINS, 101

N

NATURAL VEGETATION, 141-145
NORTHERN HEMISPHERE, 33-34
NORTHINGS, 20
NORTH POLE, 32, 33, 34, 36, 38
NUCLEATED SETTLEMENT PATTERN, 160

O

OBLIQUE MERCATOR PROJECTION, 78
OCEAN CURRENTS, 117, 121

OROGRAPHIC PRECIPITATION, 124
OUTLINE MAPS, 8-11

P
PARALLELS OF LATITUDE, 32-38
  (see latitude)
PATTERNS MADE BY RIVERS, 108
  dendritic, 108
  radial, 108
  trellis, 108
PHOTOGRAPHS, 87-94
  air photographs, 89-94
  background, 88
  foreground, 88
  ground level, 87-88
  oblique air photographs, 89-91
  patterns, 87
  types, 87-88
  vertical air photographs, 92-94
PHYSICAL FEATURES, 101-112
PHYSICAL MAPS, 69
PICTORIAL MAP, 13
PLACE NAME OR POLITICAL MAPS, 66-67
PLAINS, 101
PLATE TECTONICS, 104
PLATEAUX, 101
PLOTTING A POINT, 20-21
PLOTTING POSITION, 18-19
POINT FORM SETTLEMENT PATTERN, 159
POPULATION, 152-157
  density, 155-157
  distribution, 152-157
  factors affecting density and distribution, 152-153
    adequate precipitation, 152
    availability of power, 152
    flat land, 152
    fresh water supplies, 152
    good soil, 153
    inertia, 153
    moderate winters, 152
    resource availability, 152
    transportation, 152
  map symbol, 29
  provinces and territories of Canada, 154
PRECIPITATION, 123-127
  climate controls, 124-126
  convectional, 125
  cyclonic, 126
  isohyets, 123
  graphs, 132-135
  mean annual, 123
  orographic, 124
PRECIPITATION GRAPHS, 132-135
PRIME MERIDIAN, 39, 41, 50
PROFILES, 99-100
PROJECTIONS, MAP, 72-81

R
RADIAL DRAINAGE PATTERN, 108
RANGE
  latitude, 96
  longitude, 96
REASONS FOR THE LOCATION OF SETTLEMENTS, 162
RECENT DEPOSITS, 105
RECTANGULAR SETTLEMENT PATTERN, 160
REGIONAL TOPOGRAPHIC MAPS, 70-71
REPRESENTATIVE FRACTION, 57-58
RHUMB LINE (LOXODROME), 85-86
RICE FARMING, 150-151
RIVER SYSTEMS, 108-112
  drainage basins, 109-110
  patterns made by rivers, 108
  spillways, 110-112
ROCKS, 101-107
ROUND SETTLEMENT PATTERN, 161

S
SCALE, 6-7, 55-65
  atlas map, 60-65
  large, 59
  line, 56-57
  map, 56-60
  medium, 59
  representative fraction, 57-58
  road map, 6-7
  small, 59
  statement, 56
SEA LEVEL, 25
SEASON REVERSAL, 119
SEDIMENTARY ROCKS, 102-103, 107
SETTLEMENT GROWTH, 163-167
SHAPES, 8-11
  continents, 11
  islands of Canada, 9
  provinces of Canada, 10

SKETCH MAP, 94
SOILS, 148
SOUTH POLE, 32, 33, 34, 36, 38
SOUTHERN HEMISPHERE, 33-34
SPILLWAYS, 110-111
SPOT HEIGHT, 25
STATEMENT SCALE, 56
SYMBOLS, 12-14
    atlas map, 27-29
    conventional, 28
    line, 27, 28
    mark, 27
    pictorial, 13, 29

T
TEMPERATURE, 113-122
    climate controls, 118-122
        altitude, 119-120
        latitude, 118-119
        ocean currents, 121
        prevailing winds, 122
        water, 120-121
    interpretation, 136
    isotherms, 113-114
    graphs, 128-131
    maps, 113
    range, 115
    shape of temperature line, 130
    word description, 137
THEMATIC MAPS, 3, 68
TIME ZONES, 48-54
    boundaries, 53
    Canada, 54
    Greenwich Mean Time (GMT), 51
    International Date Line, 52

TITLE, 4
TOPOGRAPHIC MAPS, 70-71
TOWNSHIPS IN THE PRAIRIE PROVINCES, 163
TRADE WINDS, 116
TRANSVERSE MERCATOR PROJECTION, 78
TRAVEL, 169
TRELLIS DRAINAGE PATTERN, 108
TRIANGULATION STATION, 26
TROPIC OF CANCER, 36
TROPIC OF CAPRICORN, 36

U
UPLIFTED REMAINS OF ANCIENT MOUNTAIN SYSTEMS, 103-104, 106

V
VEGETATION, 141-145
VERTICAL AIR PHOTOGRAPHS, 92-94
VOLCANOES, 104

W
WARM FRONT, 126
WATERSHEDS, 109
WESTERN HEMISPHERE, 40
WHEAT FARMING, 149

Y
YOUNGER FOLD MOUNTAINS, 104-105, 107

Z
ZENITHAL EQUAL AREA PROJECTION, 81
ZENITHAL EQUIDISTANT PROJECTION, 80

ACKNOWLEDGEMENTS

The author and the publisher wish to acknowledge the kindness of those who granted permission for the reproduction of the photographs and maps indicated below. Every reasonable care has been made to find ownership of copyrighted photographs and maps. Information will be welcome which will allow the publisher to rectify any credit in subsequent editions.

Public Archives of Canada, p. 2 (lower); James Galt and Company Limited, Cheadle, Cheshire, England, p. 13; British Columbia Ministry of Agriculture and Food, p. 22 (upper); Muriel Fiona Napier, p. 22 (lower); New Brunswick Travel Bureau, p. 23; Ontario Ministry of Transportation and Communications, p. 26; Miller Services Limited, Toronto, Ontario, p. 53 (upper and lower); Ontario Ministry of Industry and Tourism, p. 88; Ontario Place Corporation, p. 90; Northway-Gestalt Corporation, Toronto, Ontario, p. 91; Northway Map Technology, p. 93; SSC — Photocentre — ASC, p. 106 (top and bottom); Northway Map Technology, p. 106 (middle); Canadian Government Travel Bureau, p. 107 (upper); SSC — Photocentre — ASC, p. 107 (lower); Ontario Ministry of Industry and Tourism, p. 143 (upper); Miller Services, p. 143 (middle and lower); United Nations, p. 144 (upper and middle); Montana Travel Promotion unit, p. 144 (lower); Australian Information Service, Ottawa, p. 145; Israel Information Service, p. 161; Courtesy of the Hispanic Society of America, New York, p. 173.